Using Classification and Regression Trees

Using Classification and Regression Trees

A Practical Primer

Xin Ma

University of Kentucky

IAP

INFORMATION AGE PUBLISHING, INC.
Charlotte, NC • www.infoagepub.com

Library of Congress Cataloging-in-Publication Data

A CIP record for this book is available from the Library of Congress
http://www.loc.gov

ISBN: 978-1-64113-237-4 (Paperback)
978-1-64113-238-1 (Hardcover)
978-1-64113-239-8 (ebook)

Printed and bound by CPI Group (UK) Ltd, Croydon, CR0 4YY

To my wife, Ping, for her years of selfless support for my academic career.

Contents

Preface

Most traditional statistical techniques rely on the development of a model to make sense of the data. When there are well-established theoretical frameworks to guide the specification of the model, a theory-driven strategy works. Otherwise, data-driven research is of great value to discover the hidden patterns and relationships in the data. This type of data analysis is computationally intensive but has become possible because of modern computing capacities. Classification and regression trees (CART) is one of the several contemporary statistical techniques with good promise for research in many academic fields.

This book is written for academic researchers, data analysts, and graduate students in fields such as economics, social sciences, medical sciences, and sport sciences who want to tap into the relatively new statistical technique of CART as a powerful analytical tool for research in their fields. This book is all about applied statistics on CART, intended to help readers become knowledgeable consumers of studies based on CART (e.g., be able to read, understand, and critique studies using CART), develop analytical skills for using CART with the assistance of statistical software programs, and venture into some advanced CART techniques currently not well-discussed in the literature (e.g., longitudinal CART analysis). High school mathematics and science teachers can also use this book as a resource for statistics courses and extracurricular activities for advanced students in mathematics and science.

Using Classification and Regression Trees, pages xi–xii

There is a limited number of books on CART, especially on applied CART. Motivated by the lack of a good practical primer about CART, this book focuses on the applications of CART, using easy (nontechnical) language and illustrative graphs (tables) as much as possible. There is no special demand for statistical software programs. In this book, the popular program, SPSS, is used to execute examples that are based on real-world data. Chapters aim to show how to formulate research questions for CART analysis, how to describe CART procedures, and how to present and interpret CART results. All these features encourage readers with minimum statistical background to become knowledgeable and skillful users of CART. The field has not yet produced any applied CART book like this one with the sole purpose of applying CART to solving real-world problems.

Though not covered in extreme detail, there are also extensions and innovations in this book that go beyond any other books on CART. This book has a chapter on some advanced CART procedures not well discussed in the literature. Yet, these advanced topics of CART are described in an easy-to-understand fashion, unintimidating to readers without strong statistical background. This feature allows them to effectively seek further empowerment of their research designs by extending the analytical power of CART to a whole new level. Overall, this innovative book provides a solid foundation for readers' exploration of CART and serves as a stepping-stone into more advanced statistical inquiry of CART.

1

Introduction

Scientific Reasoning in the Computer Age

In their article, "Statistical Data Analysis in the Computer Age," featured in *Science,* Efron and Tibshirani (1991) stated that:

> [m]ost of our familiar statistical methods, such as hypothesis testing, linear regression, analysis of variance, and maximum likelihood estimation, were designed to be implemented on mechanical calculators. Modern electronic computation has encouraged a host of new statistical methods that require fewer distributional assumptions than their predecessors and can be applied to more complicated statistical estimators. These methods allow the scientist to explore and describe data and draw valid statistical inferences without the usual concerns for mathematical tractability. This is possible because traditional methods of mathematical analysis are replaced by specially constructed computer algorithms. Mathematics has not disappeared from statistical theory. It is the main method for deciding which algorithms are correct and efficient tools for automating statistical inference. (p. 390)

This is part of a general intellectual effort to extend the ability of human proof by tapping into the power of "electronic computation" (using the

Using Classification and Regression Trees, pages 1–14

expression of Efron and Tibshirani). Such an advancement in reasoning is not without controversy.

Sara Billey (2015) of the University of Washington humorously referred to this effort as *Computer Assisted Proofs: Coming Soon to a Theorem near You.* She used as a classic example for demonstration the four-color map theorem that states that every map of counties or states or countries can be colored with 4 colors so that any two adjacent regions can have different colors. Francis Guthrie conjectured this theorem to be true in 1852. Many proofs were proposed and rejected over the years until 1976 when Kenneth Appel and Wolfgang Haken (1977) offered a computer assisted proof. To put in an oversimplified way, they exhausted all the possible discrete map scenarios using the power of a computer. The proof was highly controversial, and the *New York Times* even refused to report it. Because of the involvement of electronic computation, some mathematicians are still unwilling to accept the proof as valid even though no logical flaws have ever been detected.

The idea of borrowing the computing power of a modern computer to exhaust all possible scenarios so as to identify the best statistical inference (e.g., the model with the best model-data-fit statistic of all possible models to be built) may not be as controversial in statistics as in mathematics but can be just as powerful in statistics as in mathematics. Classification and regression trees (CART) is among the host of new statistical techniques of the computer age. Developed in 1984 by Leo Breiman and Charles Stone of the University of California at Berkeley and Jerome Friedman and Richard Olshen of Stanford University (Breiman, Friedman, Olshen, & Stone, 1984), CART is a decision-tree procedure that functions to classify cases and make predictions. One can think of a decision tree as a flow chart that shows a logical path of answers to a sequence of questions. According to its characteristics, a case can be traced down the logical path (or tree structure) to its destination where a qualitative statement and a quantitative prediction can be made about a group of cases similar to the one at hand.

CART is showing some very good promises for research in many fields. This book aims to introduce CART as a contemporary and powerful means of data analysis. Statistically speaking, CART is an exploratory procedure (i.e., exploratory data analysis) associated with what is often referred to as inductive research or data-driven research. A directly relevant but more contemporary term or concept is data mining, largely a "product" of the computer age. The online *Merriam-Webster Dictionary* defines data mining (n.d.) as "the practice of searching through large amounts of computerized data to find useful patterns or trends." This short paragraph attempts to quickly establish that CART, as a data-mining technique, is a new tool for exploratory data analysis often performed in inductive research or data-driven research.

The discussion throughout this book attempts to demonstrate that CART has a great potential to bring quantitative (inductive) research to a whole new level that was unimaginable before the computer age.

Making the Case for Inductive or Data-Driven Research

Scientific research methods can often be classified into two different types of reasoning: deductive and inductive. Simply put, reasoning is deductive when a conclusion is a logical result of a premise (e.g., a dark cloud brings rain; I see a dark cloud, it is going to rain), whereas reasoning is inductive when a premise derived from a set of specific facts leads to a general conclusion (e.g., 2 [an even number] + 3 [an odd number] = 5 [an odd number]); one gets an odd number when an even number is added to an odd number].

Over the past few decades, there has been a general academic trend in favor of deductive, or theory-driven, research; while inductive, or data-driven, research has been criticized for lack of theoretical foundations or unifying concepts that are able to anchor the research. For example, Aneshensel (2002) believes that the advent of complex and powerful computer-generated statistical techniques is a serious threat to the prominence of social theory in data analysis. She reasons that social sciences research should emphasize social theories rather than statistical techniques. The former must dictate the latter, but the opposite should be avoided. In many academic circles, one can easily hear expressions that discourage the data-driven approach of research (e.g., fishing trip and playing with data).

The current debate over theory-driven research versus data-driven research is not a new one in the research community. The competing theme of exploratory versus confirmatory data analysis has been around for decades. "Exploratory data analysis emphasizes flexible searching for clues and evidence, whereas confirmatory data analysis stresses evaluating the available evidence" (Hoaglin, Mosteller, & Tukey, 2000, p. 2). Although the merits of theory-driven research are sound and should be put into academic practice, the contribution of data-driven research is equally legitimate. Hoaglin et al. (2000) stressed that "good statistical practitioners have always looked in detail at the data before producing summary statistics and tests of hypotheses" (p. 1).

John Tukey described exploratory data analysis as "detective work" (1977, p. 1). He stated that

> [a]s all detective stories remind us, many of the circumstances surrounding a crime are accidental or misleading. Equally, many of the indications to be discerned in bodies of data are accidental or misleading. To accept all

> appearances as conclusive would be destructively foolish, either in crime detection or in data analysis. To fail to collect all appearances because some—or even most—are accidents would, however, be gross misfeasance deserving (and often receiving) appropriate punishment. (p. 3)

Tukey (1977) also discussed in several other places the case of *Exploratory Data Analysis* as he purposefully entitled his classic book:

- "Unless exploratory data analysis uncovers indications, usually quantitative ones, there is likely to be nothing for confirmatory data analysis to consider" (p. 3).
- "Restricting one's self to the planned analysis—failing to accompany it with exploration—loses sight of the most interesting results too frequently to be comfortable" (p. 3).
- "Exploratory data analysis can never be the whole story, but nothing else can serve as the foundation stone—as the first step" (p. 3).
- "Today, exploratory and confirmatory can—and should—proceed side by side" (p. vii).

Other statisticians have argued from a different perspective for the legitimacy of data-driven research. Scientific research usually unpacks relationships. In his *Wald Lecture Series*, Leo Breiman (2002) thinks of data as being generated by a black box—input (independent) variables go into one side and response (dependent) variables come out on the other side. The purpose of statistical analysis is to draw conclusions about the mechanism operating inside the black box. From a statistical perspective, it does not matter whether a statistical model comes from a deductive or inductive approach. What matters is the model-data-fit. The better the model fits the data, the sounder the inferences are about the black box (see Breiman, 2002). Following this line of logic, data-driven research is just as legitimate as theory-driven research, if one's goal is to expose the mechanism operating inside the black box.

Putting the Case in a Practical Perspective

From a practical perspective, theory-driven research relies on the specification of a statistical model based on a unifying concept or theory to describe the relationship between dependent and independent variables. This strategy works well when there is a sufficient theoretical framework or adequate previous research to guide the development of the model. It becomes problematic when there are few theories or studies to guide the specification of the model. The problem becomes particularly serious when there are a large number of independent variables in the model. In situations like this, data-driven research that is able to unpack the mechanism functioning inside the black box appears to be a better alternative than theory-driven research that has little chance to prescribe with certainty a sound statistical model.

There is no lack of examples in reality where data-driven research is instrumental. For example, school principals deal with many issues in learning and teaching as well as in operation and management that do not have adequate working knowledge behind them. In response, methods of exploratory data analysis have been included in the McNamara/Thompson guidelines for teaching basic statistics in principal preparation programs (see McNamara, 2000). For other examples, Sinacore, Chang, and Falconer (1992) used the evaluation of a rehabilitation program for people with rheumatoid arthritis as a case to argue for the benefit of applying exploratory data analysis to program evaluation research. Navigating in an under-researched area, Suyemoto and MacDonald (1996) employed a flexible, data-driven research method to derive an inductive theory concerning the content and function of religious beliefs.

It could be useful to discuss one example in greater detail. Ma (2003) investigated the training (learning) behaviors of self-employed people in Canada, noticing that "during the last few decades of the 20th century, there has been a dramatic rise in the rate of self-employment within many industrialized countries" (Hughes, 1999, p. 1). Murray and Zeesman (2001) argued that to compete effectively in the new global economy, workers must renew their skill bases and acquire new competencies. A timely and sensitive social policy issue is to identify the characteristics of self-employed people who participate in different kinds of training (learning) activities.

Among all employment strategies, self-employment is one of the least researched and understood. This situation is certainly understandable given that self-employment is a late 20th century phenomenon (in terms of scope and scale). In particular, there is a serious paucity of research on the training (learning) behaviors of self-employed people. So rare are empirical studies, theoretical thoughts, or even scientific hypotheses about the training behaviors of self-employed people. In response to the demand of policymakers and administrators for working knowledge in their effort to promote and improve self-employment, the Survey of Self-Employment ([SSE]; Human Resources Development Canada & Statistics Canada, 2000) was designed and implemented. Part of the collected data pertained to the training behaviors of self-employed people. These data have provided excellent opportunities for inductive, data-driven research in the absence of theoretical frameworks built on (previous) empirical studies.

Suppose one intends to examine the determinants of individual participation in training with six demographic variables as potential candidates (the dependent variable, participation, is dichotomous). The simplest approach is to run a series of cross tabulations to identify variables that are related to participation in training. However, without theories to guide data analysis, one is often reluctant to take this approach because it does not consider potential interactions among those variables. What about building interactions into the analysis? The idea is thoughtful but the practice is difficult. One needs to examine all two-way, three-way, four-way, five-way, and six-way tables to properly identify potential interaction effects.

To avoid tediousness and confusion, one often uses log-linear models to analyze multi-way, cross-table data. If the dependent variable has two categories and each independent variable has three categories, a log-linear model contains $2 \times 3 \times 3 \times 3 \times 3 \times 3 \times 3 = 1{,}458$ cells. It is virtually certain that many cells are empty, creating an unbalanced design which in turn biases statistical estimates. An enormous sample would be needed to fill up each cell with a sufficient number of cases. The word "sufficient" is important because cell size directly influences the statistical power of the log-linear model (see Cohen, 1988). About 50 individuals per cell is on the safe side, resulting in a sample size of $1{,}458 \times 50 = 72{,}900$. This sample size (based on six independent variables) is rarely achievable in reality, let alone studies that involve a large number of independent variables.

Because complex interactive relations among independent variables are often difficult to pinpoint (especially when there are a large number of independent variables), misspecified models usually occur in data analysis under insufficient guidance from theoretical frameworks. In cases like the one above, data-driven statistical procedures such as CART are appropriate to discover the latent patterns of relations in the data, which could help set the foundation for theoretical development.

Demonstration of CART as an Exploratory Technique

Leaving the specific CART procedures for later discussion, Tables 1.1 and 1.2 present the CART tree structure (in table format) on the self-employed using both formal and informal training (learning).[1] Formal training is the traditional way of delivering knowledge in the form of courses (long or short) through certain methods (in-person, long-distance, or web-based). Informal training in which individuals upgrade their work skills through various informal channels such as discussions, demonstrations, and conferences has started to draw attention from policymakers. Table 1.1 shows that the CART tree

TABLE 1.1 Participation Rate in Percentage of Canadian Self-Employed Using Both Formal and Informal Training

Terminal Group	Terminal Size	Participation Rate	Percent Index
Group 1	188	3.2	0.125
Group 2	896	11.3	0.441
Group 3	680	16.5	0.645
Group 4	405	22.0	0.860
Group 5	164	20.1	0.788
Group 6	79	6.3	0.248
Group 7	136	11.0	0.432
Group 8	192	31.3	1.223
Group 9	83	56.6	2.217
Group 10	271	29.5	1.156
Group 11	309	47.9	1.875
Group 12	207	66.2	2.591
Group 13	74	47.3	1.851
Group 14	156	72.4	2.835
Sample	3,840	25.6	

contains 14 mutually exclusive terminal groups (that cannot be divided any further). Although terminal groups vary in size, individuals in each terminal group demonstrate similar training behaviors. The participation rate indicates the percentage of individuals who engage in both formal and informal training in a certain terminal group. The percent index is calculated as the ratio between the group average participation rate and the sample average participation rate. Thus, groups whose average participation rates are larger than the sample average participation rate have indices above 100; and the larger the index, the more likely for individuals in the associated group to use both formal and informal training. In contrast, groups whose average participation rates are smaller than the sample average participation rate have indices below 100; and the smaller the index, the less likely for individuals in the associated group to use both formal and informal training.

The interpretation of Table 1.1 would come after the introduction of Table 1.2. Although information from both tables should be examined together to make sense (i.e., be meaningful), it is simply impossible to merge the two tables together. One common challenge for researchers using CART is to develop efficient and effective ways to present or report the CART results which tend to be massive. Table 1.2 describes the characteristics of individuals in each terminal group. Independent variables (as predictors)

TABLE 1.2 Classification (Profile) of Canadian Self-Employed Using Both Formal and Informal Training

Terminal Group	First Significant Predictor	Second Significant Predictor	Third Significant Predictor
Group 1	Does not belong to an association	0 to 8 years of education	
Group 2	Does not belong to an association	Some secondary, grades 11 to 13	
Group 3	Does not belong to an association	Some post secondary or higher	Does not have employees
Group 4	Does not belong to an association	Some post secondary or higher	Have employees
Group 5	Belong to an association	0 to 8 years, some secondary	Work less than 50 hours per week
Group 6	Belong to an association	0 to 8 years, some secondary	Work more than 50 hours per week
Group 7	Belong to an association	Grades 11 to 13, some post secondary	Job in 2, 3, 4, 6, 8, 15
Group 8	Belong to an association	Grades 11 to 13, some post secondary	Job in 1, 5, 7, 12, 14, 16
Group 9	Belong to an association	Grades 11 to 13, some post secondary	Job in 9, 10, 11, 13
Group 10	Belong to an association	Diploma, bachelors degree	Job in 1, 3, 4, 5, 6, 8, 15
Group 11	Belong to an association	Diploma, bachelors degree	Job in 7, 10, 11, 16
Group 12	Belong to an association	Diploma, bachelors degree	Job in 2, 9, 12, 13, 14
Group 13	Belong to an association	Graduate degree	Does not have employees
Group 14	Belong to an association	Graduate degree	Have employees

Note: Industry of main job current or held in the past year includes 1 = agriculture; 2 = forestry, fishing, mining, and oil; 3 = construction; 4 = manufacturing (durables); 5 = manufacturing (nondurables); 6 = wholesale; 7 = retail trade; 8 = transportation and warehousing; 9 = finance, insurance, and real estate; 10 = professional, scientific, and technical; 11 = management and administrative support; 12 = educational services; 13 = health care and social assistance; 14 = information, culture, and recreation; 15 = accommodation and food services; and 16 = other services. The first significant predictor is association membership ($\chi^2 = 337.40$, $df=1$). The second significant predictor is education (appearing twice) ($\chi^2 = 41.66$, $df = 2$ and $\chi^2 = 148.08$, $df = 3$). The third significant predictors are employees in the past year (appearing twice) ($\chi^2 = 5.10$, $df = 1$ and $\chi^2 = 13.83$, $df = 1$), actual hours per week at all jobs ($\chi^2 = 7.69$, $df = 1$), and industry of main job (appearing twice) ($\chi^2 = 51.77$, $df = 2$ and $\chi^2 = 63.90$, $df = 2$).

pertain to demographic characteristics of the self-employed, including gender, age, marital status, immigration, family size, education (level), and earnings of self-employment, as well as environmental characteristics, including city of residence, region of residence, whether spouse is a business partner, whether self-employment is incorporated, whether employees are hired, how previous employment ends, membership to a professional association, previous experience in self-employment, total work hours per week, area of self-employment (main job), and years in self-employment. These demographic and environmental characteristics are explored, by means of CART, for their relationships with the training (learning) behaviors of the self-employed (i.e., whether or not the self-employed take part in both formal and informal training).

Tables 1.1 and 1.2 contain a large amount of information. One may want to focus on the terminal groups with high and low probabilities of using both formal and informal training (note that participation rates in Table 1.1 can be understood as simple probabilities). Often times, one compares the participation rate of a terminal group with the average participation rate of the sample to gain some understanding about individual behaviors in that particular terminal group. Obviously, the percent index fulfills this idea (see Table 1.1).

The terminal group with the highest participation rate is the 14th group with 156 individuals (equivalent to 156 ÷ 3,840 = 4.1% of the sample). This group consists of the self-employed who belong to professional associations, who have graduate degrees, and who have employees (see Table 1.2). Among individuals in this group, 72.4% have used both formal and informal training, in comparison to the sample average participation rate of 25.6%. The percent index for this group is 2.828 (calculated as 72.4 ÷ 25.6), indicating that its participation rate is almost 3 times higher than the average participation rate of the sample.

The terminal group with the lowest participation rate is the first group with 188 individuals (equivalent to 188 ÷ 3,840 = 4.9% of the sample). This group consists of the self-employed who do not belong to professional associations and who have less than 8 years of education. Among individuals in this group, 3.2% have used both formal and informal training, in comparison to the sample average participation rate of 25.6%. The percent index for this group is 0.125 (calculated as 3.2 ÷ 25.6), indicating that its participation rate is 8 times (equivalent to the reciprocal of 0.125) lower than the average participation rate of the sample.

Although these two groups provide insightful information, they are somewhat extreme given their small sizes in the sample. Some may be more interested in the most "populated" terminal group. That group is the second one with 896 individuals (equivalent to 896 ÷ 3,840 = 23.3% of the sample). This group consists of the self-employed who do not belong to professional associations and who have some secondary school education. Among individuals in this group, 11.3% have used both formal and informal training, in comparison to the sample average participation rate of 25.6%. The percent index for this group is 0.441 (calculated as 11.3 ÷ 25.6), indicating that its participation rate is more than twice (equivalent to the reciprocal of 0.441) as low as the average participation rate of the sample.

To some, these interpretations sound much like profiles. As a matter of fact, profiling terminal groups is a very powerful function of CART. Because individuals are classified into mutually exclusive terminal groups, questions can be asked about who these individuals are in a certain terminal group. When individual characteristics of a terminal group are paired with the value on the dependent variable for that terminal group, one gains a good insight into a unique group of individuals. When profiles are established, predications of many kinds can be made for a variety of research and policy purposes. For example, in the current case, to improve training of the self-employed, one of the targets is those who do not belong to professional associations and who have no further education beyond secondary schooling (the most populated terminal group but with a problematic participation rate in training).

Apart from specific information on terminal groups, a general question can also be asked regarding what independent variables are, overall, significant to the training behaviors of the self-employed.[2] When demographic and environmental variables are considered together, the first significant predictor, association membership, comes from the set of environmental variables (see Table 1.2). The second significant predictor, education, comes from the set of demographic variables. All of the third significant predictors come from the set of environmental variables (employees in the past year, actual hours per week at all jobs, and industry of main job current or held in the past year). Therefore, the results indicate that environmental characteristics of the self-employed are more important than demographic characteristics to their engagement in both formal and informal training.

What makes this analysis unique is that it is able to uncover the group dynamics in the training behaviors of the self-employed. Group dynamics refers to the capability of CART to channel individuals into dramatically different groups in terms of the use of both formal and informal training. Note that the participation rates of the terminal groups vary from 3.2 to 72.4%. This group dynamics comes to light as a result of decomposition of interaction effects among independent variables. Now one may wonder where the interactions are in Tables 1.1 and 1.2. Indeed, the interactions are not obvious in those tables because they are a narrative summary of the CART tree. The graphic illustration of the CART tree is far more revealing of how the interactions "grow" the tree. To avoid the confusion potentially caused by a whole (big) tree, Figure 1.1 presents a small portion of it just for demonstration purposes.[3] One needs to have more than one variable to discuss any interactions. The terminal groups are not a product of one variable but at least two variables in Figure 1.1. A specific category of association membership needs to work with a specific category of education to produce a terminal group—the very essence of interactions. For example, when no association membership pairs with education of 8 years or less, a unique terminal group is formed—the interaction grows the tree. Similarly, a specific category of education needs to work with a specific category of employees to produce a terminal group. For example, when some post secondary education pairs with no employees, a unique terminal group is formed. Again, the interaction grows the tree. To a large extent, the CART tree itself is simply a graph of interactions, in fact, nothing but interactions.

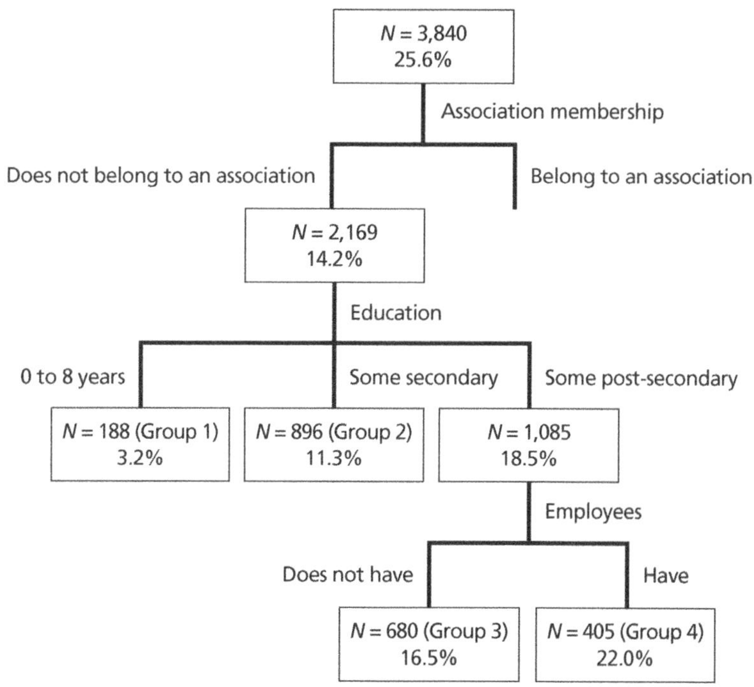

Figure 1.1 Partial CART tree on training behaviors of the self-employed. In each box, the top number indicates the number of individuals and the bottom number indicates the average participation rate of individuals in training.

Coming back to the issue of group dynamics revealed in the CART analysis, Table 1.2 offers insight into the mechanism operating inside of the black box through the unearthing of the interactions. Again, individuals in each terminal group can be fully described of their most important demographic and environmental characteristics (as a result of interactions among predictors). Based on this inductive research, theories or at least hypotheses can now be formed for later deductive research (as advocated in the writings of John Tukey and others). Analysis associated with Tables 1.1 and 1.2 is a good demonstration of the analytical and exploratory power of CART—the new generation of statistical techniques—that is capable of disentangling complex interactions among the independent variables. As a result, CART enables one to detect the group dynamics in a way that is usually impossible with traditional statistical techniques.

Advantages of CART

As a family of advanced statistical techniques, CART clusters individuals into a number of mutually exclusive and exhaustive groups based on interaction effects among the independent variables. CART is an effective exploratory statistical technique. The statistical principle of CART can be summarized as recursive partitioning; that is, progressively dividing individuals into smaller and smaller groups with increasing similarities in the dependent variable within each group and meanwhile with increasing differences in the dependent measure between newly formed groups. CART has several advantages over traditional statistical techniques (see Clarke, Bloch, Danoff, & Esdaile, 1994).

First, as discussed above, most traditional statistical techniques rely on the development of a statistical model to describe the relationship between dependent and independent variables. Difficulties in pinpointing complex interactions among independent variables frequently result in mis-specified models. Prior identification and modeling of interactions are not necessary in CART because it automatically generates mutually exclusive groups that provide direct insight into the interactive nature of significant independent variables. Therefore, CART can capture complex interactions and nonlinear relationships in the data with which traditional statistical techniques cannot easily deal.

Second, without relying on any statistical model, CART does not contain complex mathematical equations (that describe the statistical model). Its results are easy to interpret and understand. Specifically, CART interpretation focuses on each terminal group of individuals whose characteristics (on the independent variables) can be fully described and on the average estimate on the dependent variable that each terminal group indicates. Generally, the tree structure automatically indicates the most important independent variables and how they interact with one another to channel individuals into markedly different groups as far as the dependent measure is concerned.

Third, traditional statistical techniques often require some distributional assumptions (e.g., normal distribution) that are usually difficult to meet in real data situations. In the example presented earlier, even if one has all the resources required to construct an enormous sample, there is no guarantee that distributional assumptions can all be met. CART, on the other hand, is a nonparametric statistical technique, free from any distributional assumptions (see Zhang & Singer, 1999).[4] Therefore, one can use CART to investigate many data sets that are traditionally considered unfruitful or inappropriate for statistical data analysis due to abnormal distributions of data.

Finally, some recent CART applications have shown potential of CART in boosting or improving the performance of traditional statistical techniques. The basic idea is to use the information that CART generates to guide the specification of traditional statistical models (e.g., multiple regression). For example, one can use CART to improve the predictive performance of multiple regression analysis by as high as 120% (see Srivastava, 2013).

Notes

1. Technically, the procedure employed in this demonstration is CHAID (chi-square automatic interaction detector). Both CART and CHAID are tree-based classification procedures. They differ only in that CART performs binary splits and CHAID performs multiple splits (see the discussion in Chapter 6). To avoid unnecessary distraction, CART is used as the "name" of the procedure in this section to maintain consistency of expression throughout this chapter.
2. Some caution is needed to refer association membership as "the first significant predictor." As a matter of fact, determining which independent variables are most important is a very difficult issue. The importance of independent variables does not depend on the order in which they appear in the tree structure. Chapter 4 discusses this issue in more detail. The criterion used to decide the first, second, and third significant variables is which independent variable can partition the entire group into two smaller groups that are as homogenous as possible within each group and as heterogeneous as possible between groups in terms of the dependent measure. The first, second, and third significant variables are defined in this sense in the analysis. This is different from the traditional notion of statistical significance which is often determined by at least two things. One is the p value (note that a variable with a certain p value can be

significant if the significance level, alpha, is set to be 0.05 but can be insignificant if the significance level is set to be 0.01); and the other is effect size which invokes some sort of standardization of effects so that the importance of each variable is brought onto the same "scale" for comparison. For this reason, some researchers applying CART prefer to use the term "the best predictor" (Ture, Kurt, Kurum, & Ozdamar, 2005, p. 584) rather than the significant predictor to indicate the difference between importance under CART and importance under traditional statistics such as multiple regression analysis.

3. In Figure 1.1, the whole sample (the very top box) is partitioned according to association membership into two branches, and Figure 1.1 represents actually the branch labeled as "Does not belong to an association." The top box of this branch (with 2,169 individuals) is partitioned according to education into three groups, two of which are terminal groups without any further partition (see Groups 1 and 2 in Tables 1.1 and 1.2). One of the three is a "transitional" group which is further partitioned into two terminal groups (see Groups 3 and 4 in Tables 1.1 and 1.2).
4. Although there is no argument from credible sources (e.g., top academic journals) in the literature against the claim that CART is non-parametric, not all statisticians accept the claim. There are two ways to approach the issue of parametric versus nonparametric. One way is to examine the existence of a probability density function (PDF). Parametric models have a PDF, whereas nonparametric models do not. For example, a normal distribution has a PDF and all models based on the normal distribution assume the PDF. The other way is to consider whether the number of parameters is fixed or not. Parametric models have a fixed number of parameters (as a result of the model specification or the research design), whereas nonparametric models do not. For example, CART does not operate on a pre-determined number of parameters or nodes (i.e., boxes in Figure 1.1); instead, CART minimizes error in the search for the best number of nodes. From this perspective, CART is definitely nonparametric. The disagreement likely comes from the PDF perspective.

2

Statistical Principles of CART

The theoretical foundations and practical applications of CART were first presented in Breiman, Friedman, Olshen, and Stone (1984). Since then, with the increasing power of computers, CART has rapidly developed into a powerful exploratory method for data analysis. Although many statisticians use the term CART as if it were one statistical method, CART actually includes two analytic methods—classification trees (CT) and regression trees (RT), depending on the measurement nature of the dependent variable. One uses CT on nominal (or categorical) dependent measures, whereas one uses RT on interval (or continuous) dependent measures. The reason why CART is often used as a general analytic term is probably that CT and RT share common statistical principles and differ only in details.

This chapter focuses on statistical issues that are common to CT and RT, using CART as a general expression. The next chapter focuses on statistical procedures that are unique to CT and RT. Specifically, this chapter introduces three measures of impurity (i.e., the degree to which cases in a group belong to different categories or values of the dependent variable) as the very foundation for CART analysis, the idea of using impurity to grow a

Using Classification and Regression Trees, pages 15–38

CART tree, three rules to stop the tree growth, and the idea of using impurity to prune a CART tree.

Important Functions of CART

In general, CART is a heuristic tree method that unpacks the relationships between an outcome measure (a dependent variable) and a group of predictors (independent variables). One can use CART to perform several analytical functions including segmentation, stratification, prediction, interaction identification, variable screening, and variable manipulation. These functions are summarized in the following space, accompanied by some examples of classic applications as demonstrations.

Segmentation aims to identify cases that are likely to belong to a certain group. Morwitz and Schmittlein (1992) investigated the use of segmentation as a way to increase the accuracy of sales forecasts based on stated purchase intent of consumers. After comparing several statistical techniques with the analytical function of segmentation, they concluded that more accurate sales forecasts can be achieved by applying CART than traditional statistical techniques such as cluster analysis. For a given level of purchase intent, CART produces meaningful, identifiable segments with varying subsequent purchase rates. The CART results directly identify consumer segments that are most likely to fulfill their purchase intentions.

Stratification aims to assign cases to various categories. Diercks et al. (2008) postulated that gender differences in risk profiles (i.e., blood urea nitrogen, systolic blood pressure, and serum creatinine) may limit the performance of available stratification algorithms for heart failure in women. CART is employed to evaluate risk stratification. Even though statistically significant gender differences are present in all three risk variables, CART effectively stratifies both genders into distinct groups with no significant difference in mortality by gender within stratified groups. Diercks et al. (2008) concluded that, regardless of gender, CART is effective at predicting risk of heart failure.

Prediction aims to create rules for the purpose of predicting future events. One important aspect of administering self-managed large storage infrastructures is to determine which data sets to store on which devices. Wang, Au, Ailamaki, Brockwell, Faloutsos, and Ganger (2004) explored the application of CART to predict the performance of a self-managed storage system as a function of input workloads. CART is used to predict from workload characteristics response times and aggregate values of the system. Wang et al. (2004) reported that CART provides reasonably accurate predictive models.

Interaction identification aims to identify relationships that define together a certain group with a unique value on the dependent variable. Breiman et al. (1984) made the detection of interaction structures the central issue of CART. Names of algorithms, such as automatic interaction detection ([AID]; Morgan & Sonquist, 1963), suggest the importance of this tree function. Based on data from women enrolled in a population-based study of subarachnoid hemorrhage, Nelson (1998) used CART to identify three main risk groups: nonsmoking elderly women with long-standing hypertension, middle-aged women of both cigarette smokers and binge drinkers, and cigarette-smoking women of relative estrogen deficiency. Nelson (1998) realized that CART not only identifies groups with varying risks but also uncovers interactions between variables that can be overlooked in the traditional application of logistic regression.

Variable screening aims to identify a small number of predictors from a large number of variables often for the purpose of building parametric statistical models. Morrison (1998) stated that "CART can also be used in regression models to add insight into what to include as explanatory factors from a large set of independent variables" (p. 12). High energy physics experiments typically generate data on a large set of variables. Comparing different statistical methods for selecting the most discriminating variables from experimental data, Proriol (1994) concluded that CART is among the best ones, still standing out with a unique advantage of faster computation.

Variable manipulation aims to collapse predictor categories and continuous variables with minimal loss of information. Working with the intrusion detection systems for computer security that examine all data features to detect intrusion or misuse patterns, Chebrolu, Abraham, and Thomas (2005) noticed that some data features are redundant or contribute little to the detection process. CART is used to streamline important input features in their attempt to build a system that is computationally efficient and effective.

These important functions make CART useful in many analytical fields such as market analysis (e.g., developing direct mailing systems that maximize response rates, identifying consumer and environment factors that influence commercial sales), credit management (e.g., using credit histories to make credit decisions), policy analysis (e.g., using screening rules to streamline hiring processes, selecting the most important variables from survey studies to inform policy and practice), and quality control (e.g., identifying efficient procedures that effectively detect product defects). Interestingly, all these functions occur in a single CART analysis. One simply takes the CART results from a unique perspective or interprets the CART tree with a specific purpose.

Statistical Concepts of CART

Figure 2.1 presents a tree generated from a CART analysis of the relationship between smoking and stress among Canadian students in grades 6 to 10. This sample (N = 11,256) comes from the Health Behaviors in School-Aged Children (HBSC), a multinational survey (see Currie, Hurrelmann, Settertobulte, Smith, & Todd, 2000) that attempts to understand the impact of family and school experiences on health outcomes and behaviors of young adolescents. The dependent variable, smoking, measures whether students have smoked at least once. There are two independent variables describing stress of students: parent (home) related stress and teacher (school) related stress. A student's gender and age as well as the number of parents who live together with the student are used as individual background variables. This analysis attempts to test the research hypothesis that stress is related to smoking behaviors among young adolescents.

The box on the top of the tree represents the entire sample of 11,256 students. By the time of the survey, 45.42% of them had smoked at least

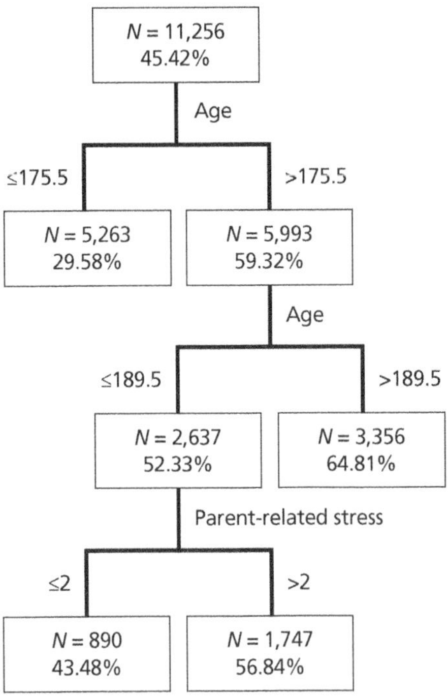

Figure 2.1 CART tree of smoking in relation to stress and background. In each node, the top number indicates the number of students and the bottom number indicates the proportion of smoking students (or the probability of smoking).

once. Among all independent variables, age (measured in the number of months) is most strongly related to smoking. Students are partitioned into two subgroups according to their age. This partition is represented by the two boxes underneath the top box. On one hand, students younger than or equal to 175.5 months form one subgroup. There are 5,263 students in this subgroup, and 29.58% of them smoked at least once. On the other hand, students older than 175.5 months form the other subgroup. There are 5,993 students in this subgroup, and 59.32% of them smoked at least once.

A further partition within the second subgroup reveals more specific age effects on smoking. Among students older than 175.5 months, those younger than or equal to 189.5 months form one subgroup where 52.33% of them smoked at least once, and those older than 189.5 months form the other subgroup where 64.81% of them smoked at least once. Students younger than or equal to 189.5 months can be further broken down into two subgroups according to their parent related stress which is measured on a five-point scale with a higher value indicating more stress. Among students younger than or equal to 189.5 months (but older than 175.5 months), those perceiving less parent related stress (scoring 1 and 2 in the five-point stress scale) form one subgroup where 43.48% of them smoked at least once, and those perceiving more parent related stress (scoring 3 to 5 in the five-point stress scale) form the other subgroup where 56.84% of them smoked at least once.

In the terminology of CART, each box (representing a group or subgroup) is called a node. The node on the top of a tree is called the root node because the analysis descends from this node. The CART analysis or the CART tree is full of partitions at different branches or at different levels. Partition refers to the splitting of cases in a node into groups. When a partition is made in CART, one node produces two consequent nodes. The produced nodes are called child nodes, whereas the producing node is called the parent node. One can distinguish the two child nodes by their positions underneath the parent node (i.e., left child node and right child node). The node that cannot be further partitioned into child nodes marks the end of growth in that part of the tree and is called a terminal node. When a root node or a parent node produces child nodes, the tree grows one level.

In Figure 2.1, the root node is the entire sample of young adolescents (11,256 students, of which 45.42% smoked). This root node is the parent node of two child nodes based on age. One of them (students younger than or equal to 175.5 months, the left child node) is a terminal node (5,263 students, of which 29.58% smoked), and the other (students older than 175.5 months, the right child node) becomes the parent node of two age-based child nodes. One of them (students older than 189.5 months) is a terminal node (3,356 students, of which 64.81% smoked), and the other (students younger than or equal to 189.5 months) becomes the parent node of two child nodes based on stress (associated with parents). Both child nodes are terminal ones (the left terminal node with 890 students, of which 43.48% smoked and the right terminal node with 1,747 students, of which 56.84% smoked). Structurally, this tree has three levels from the root node.

Although the tree structure in a CART analysis is informative, showing the interactions among the independent variables in relation to the dependent variable, the focus of the CART interpretation is often on the terminal nodes. These terminal nodes often show dramatically different outcomes on the dependent variable. In the current example, the four terminal nodes have quite different percentages (indicating probabilities) of smoking, ranging from 29.58% to 64.81%. Students younger than or equal to 179.5 months are the least likely group to smoke, whereas students older than 189.5 months are the most likely group to smoke. Tracing backward from a terminal node to the root node allows one to adequately describe the key characteristics of that terminal node. To provide fuller insights, one often calculates the mean values on each of the independent variables associated with each of the terminal nodes (see Table 2.1).

In general, age appears to be strongly related to smoking. Parent related stress indeed turns out to be a factor related to smoking, but stress has effects only within a certain age group. Among students aged between 179.5 and 189.5 months, higher parent related stress is associated with higher likelihood to smoke. Absent in the tree, teacher (school) related stress is not much associated with smoking. Overall, analytical results indicate four things associated with stress (the focus of the analysis). First, secondary to age, stress is not the strongest factor for smoking. Second,

TABLE 2.1 Stress and Background Characteristics for Each Terminal Group in the CART Tree

	Group 0	Group 1	Group 2	Group 3	Group 4
Parent Related Stress (Scale of 1–5)	2.91	2.72	3.09	1.64	3.80
Teacher Related Stress (Scale of 1–5)	2.48	2.31	2.67	2.41	2.67
Gender (Proportion of Male Students)	0.52	0.52	0.51	0.57	0.51
Age (132–241 Months)	177.11	160.25	199.26	182.43	182.80
Number of Parents (0–2)	1.79	1.79	1.80	1.80	1.80

Note: Group 0 represents the root node. Group 1 represents the terminal node at the first level. Group 2 represents the terminal node at the second level. Groups 3 and 4 represent the terminal nodes at the third level (from left to right).

there is no comprehensive impact of stress on smoking, given that only parent related stress has effects on smoking. Third, (parent related) stress is an issue only within a certain (age-based) group of students. Finally, the terminal nodes produced by stress do not indicate the highest likelihood to smoke. Therefore, analytic results seem to indicate a fairly limited impact of stress associated with families and schools on smoking behaviors, especially in the presence of student background variables.

This simple analysis shows that CART does have a powerful ability to channel students into terminal nodes with dramatically variable outcomes on the dependent measure. In addition, the partition of age within age (see the growth of the tree from the first level to the second level in Figure 2.1) is a good analytic function not readily achievable in traditional statistical techniques such as analysis of variance and multiple regression analysis. Finally, the local impact of parent related stress on smoking as shown in the CART analysis (see the third level of the tree) actually indicates a local interaction of parent related stress with age. The term "local" indicates that parent related stress does not interact with the whole spectrum of age (between 132 and 241 months in the current example) as typically seen in analysis of variance or multiple regression analysis. Instead, parent related stress interacts locally with a specific subrange of age (between 179.5 and 189.5 months in the current example). This kind of local interaction is often difficult to pinpoint with traditional statistical techniques.

When a tree generated from a CART analysis is simple, a tree diagram (as in the current example) is an effective way to illustrate the relationship between the dependent and independent variables. In contrast, the example presented in the previous chapter employs a table diagram, because the analytic results are complex with many tree branches. The decision to choose between a tree diagram and a table diagram is often based on which diagram makes it easier for researchers to interpret the results and for research information consumers (i.e., readers) to understand the results.

Statistical Procedures of CART

Statistically, CART performs successive binary partitions (splits) of groups at each level when a tree grows. The first question that many CART learners ask is why CART performs binary partitions. Theoretically, one can set the number of partitions at a parent node and even make the number of partitions vary throughout a tree (the number of child nodes descending from a parent node is often referred to as the branching factor). Note that when a parent node produces child nodes, it produces at least two child nodes (see Figure 2.1). In other words, every parent node can surely be partitioned into two child nodes. On the other hand, there is no guarantee that a parent node is able to descend, say, four child nodes, as one may prescribe. Therefore, many statisticians believe that binary partition is not only universally expressive but also comparatively simple to interpret and understand. As a matter of fact, binary partition can build any possible tree.[1] As a result, it has become a statistical convention to have CART perform binary partitions when growing a tree.

For CART, the partitioning (splitting) of cases into groups at each level is guided not by any statistical test but by a statistical criterion referred to as impurity (see Breiman et al., 1984). Impurity measures the degree to which cases in a group belong to different categories (values) of the dependent variable. A group is called pure when all cases in that group belong to a single category (or value) of the dependent variable, whereas a group is called impure when an equal number of cases belong to different categories (values) of the dependent variable. Many other (impure) situations fall between these two extremes. When a group is pure, one can terminate that part of the tree (the group becomes a terminal node). When a group is impure, one needs to decide either to stop partitioning and accept the group as a terminal node (an imperfect decision obviously) or to select another independent variable to grow the tree further until each of the child nodes is pure. Because partitions are split into sub-partitions (i.e., nodes are split into sub-nodes), CART is a recursive tree-growing process (see Lewis, 2000).

"The fundamental principle underlying tree creation is that of simplicity: We prefer decisions [partitions] that lead to a simple, compact tree with few nodes" (Duda, Hart, & Stork, 2001, p. 398). This principle reflects the philosophical notion often referred to as Occam's razor—the simplest model that explains the data is the best model. The application of this principle to CART is to seek an independent variable at each parent node that produces child nodes as pure as possible. However, rather than working with the purity of a node, it is usually mathematically more convenient to work with the impurity of the node.

Although impurity can be conceptually defined in different ways, all measures of impurity share the same behavior. The impurity of a node is zero if all cases in that node belong to a single category of the dependent variable, and impurity becomes large if an equal number of cases belong to different categories of the dependent variable. One popular impurity measure is the entropy impurity (Breiman et al., 1984)

$$i(\tau) = -\sum P(c_j)\log P(c_j)$$

TABLE 2.2 Data to Calculate Entropy Impurity Measures for Parent and Child Nodes

	Partitioning	Non-Smoking	Smoking	Row Total
Left Child Node (τ_L)	Stress ≤ 2	503 (n_{11})	387 (n_{12})	890 ($n_{1\bullet}$)
Right Child Node (τ_R)	Stress > 2	754 (n_{21})	993 (n_{22})	1,747 ($n_{2\bullet}$)
Parent Node (τ)	—	1,257 ($n_{\bullet 1}$)	1,380 ($n_{\bullet 2}$)	2,637 ($n_{\bullet\bullet}$)

Note: The parent node represents students who are older than 175.5 months but younger than or equal to 189.5 months.

where $P(c_j)$ represents the probability that a case falls into the category c_j or the proportion of the cases that go into that category in node τ. Logarithm is base 2.

To understand how this impurity measure works, consider the data in Table 2.2 which is a detailed breakdown of students at the third (bottom) level of the CART tree in Figure 2.1. The entropy impurity for the left child node can be written as

$$i(\tau_L) = -\frac{n_{11}}{n_{1\bullet}}\log\left(\frac{n_{11}}{n_{1\bullet}}\right) - \frac{n_{12}}{n_{1\bullet}}\log\left(\frac{n_{12}}{n_{1\bullet}}\right)$$

and that for the right child node can be written as

$$i(\tau_R) = -\frac{n_{21}}{n_{2\bullet}}\log\left(\frac{n_{21}}{n_{2\bullet}}\right) - \frac{n_{22}}{n_{2\bullet}}\log\left(\frac{n_{22}}{n_{2\bullet}}\right)$$

where τ represents the parent node that descends the left child node τ_L and the right child node τ_R.

Substituting numbers from the left child node in Table 2.2 produces the entropy impurity for that node

$$i(\tau_L) = -(503/890)\log(503/890) - (387/890)\log(387/890) = 0.9877\,.$$

The entropy impurity for the right child node can be calculated in the same manner

$$i(\tau_R) = -(754/1{,}747)\log(754/1{,}747) - (993/1{,}747)\log(993/1{,}747) = 0.9865.$$

The other popular impurity measure is the Gini measure of dispersion (Breiman et al., 1984). Using $P(c_j)$ to denote the percentage of the cases belonging to the category c_j (of the dependent variable) in node τ, the Gini measure is

$$i(\tau) = 1 - \sum P(c_j)^2.$$

Consider again the data in Table 2.2. The dependent variable is dichotomous with two categories (nonsmoking and smoking). In the left child node, the distribution of percentages of students falling into each category is (0.5652, 0.4348). That is

$$\frac{503}{890} \approx 0.5652 \text{ and } \frac{387}{890} \approx 0.4348.$$

The Gini measure is then

$$i(\tau_L) = 1 - (0.5652^2 + 0.4348^2) = 0.4915.$$

The Gini measure for the right child node can be calculated in the same manner

$$i(\tau_R) = 1 - (0.4316^2 + 0.5684^2) = 0.4906.$$

Still, there is another popular impurity measure called the misclassification impurity (Breiman et al., 1984) which is defined as

$$i(\tau) = 1 - \max P(c_j).$$

This definition indicates that the misclassification impurity measures the minimum probability that a case can be misclassified at the node τ. Consider the left child node in Table 2.2 in which the percentages of students classified into each category are 0.5652 (for nonsmoking) and 0.4348 (for smoking). When the dependent variable has two categories, obviously the misclassification impurity is the smaller of the two percentages

$$i(\tau) = 1 - \max(0.5652, 0.4348) = 1 - 0.5652 = 0.4348.$$

These impurity measures rarely produce different results in a CART analysis. Duda et al. (2001) stated that "an entropy impurity is frequently used because of its computational simplicity and basis in information theory, though the Gini impurity has received significant attention as well" (p. 401). Figure 2.2 compares the behaviors of the three impurity measures (entropy, Gini, and misclassification) in the case of partition into two categories (i.e., binary partition). The horizontal axis represents the percentage or proportion of cases that goes into one category or the probability of

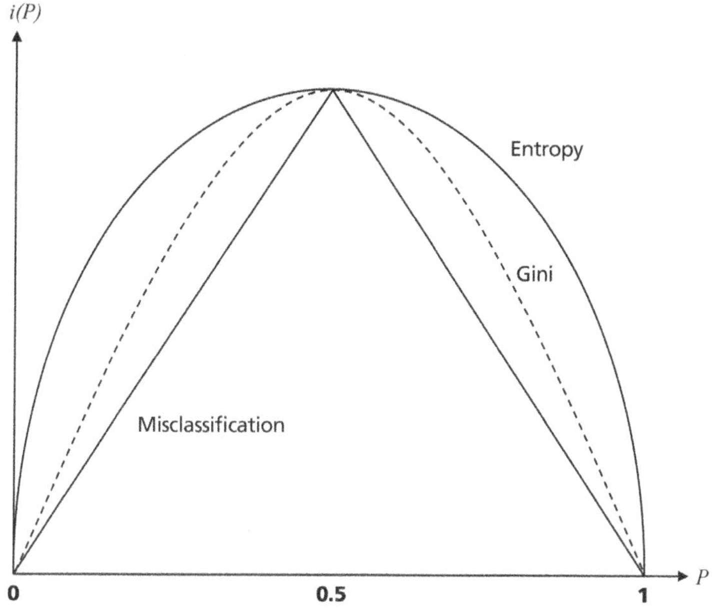

Figure 2.2 Scale simplified impurity functions in the case of binary partition (into two categories).

a case that goes into that category, and the vertical axis represents the impurity. In the figure, all impurity measures peak at the 50–50 split (partition), the situation where an equal number of cases goes to the two categories. The graph is symmetrical because, say, the 30–70 split is by nature the same as the 70–30 split in terms of impurity. For the same split, entropy yields the largest impurity value.

Impurity can also be defined for a branch (or even a tree). The idea is to calculate the weighted average of the impurity values from the child nodes forming the branch. The proportions of cases in partitioned (child) nodes are often used as weights. Again, consider the data in Table 2.2 in which the parent node is partitioned into two child nodes. Of the total of 2,637 students, 34% fall into the left child node and 66% fall into the right child node. Given their Gini impurity measures of 0.4915 and 0.4906 respectively, the impurity measure for this branch is calculated as

$$i(\tau_L, \tau_R) = 0.34 \times 0.4915 + 0.66 \times 0.4906 = 0.4909$$

where $i(\tau_L, \tau_L)$ represents the Gini impurity measure for the branch made of τ_L and τ_R.

Growing the CART Tree

CART grows a tree using the reduction in impurity as the guideline. Starting from the root node, each independent variable is examined as a potential candidate to partition the root node into two child nodes. Different categories (or values) of an independent variable can be used to partition the root node, and the optimal performance (the best reduction in impurity associated with a particular category or value) of the independent variable is recorded. The degree of reduction in impurity associated with partitioning a parent node into two child nodes is calculated as

$$\Delta = i(\tau) - i(\tau_L)\frac{n_{1\cdot}}{n_{1\cdot}+n_{2\cdot}} - i(\tau_R)\frac{n_{2\cdot}}{n_{1\cdot}+n_{2\cdot}}$$

where $i(\tau)$ is the impurity for the parent node. The coefficients associated with the child nodes can be generally considered the probabilities that a case goes into τ_L and τ_R respectively.

Using the entropy impurity as an example, the impurity of the parent node in Table 2.2 is calculated as

$$i(\tau) = -(1{,}257/2{,}637)\log(1{,}257/2{,}637) - (1{,}380/2{,}637)\log(1{,}380/2{,}637)$$
$$= 0.9984.$$

Knowing the entropy impurity measures for the parent and child nodes, one can easily calculate the reduction in impurity associated with the partitioning of the parent node into the child nodes

$$\Delta = 0.9984 - 0.9877 \times 0.34 - 0.9865 \times 0.66 = 0.0115.$$

Obviously, using a different stress value to partition the parent node results in a different reduction in the entropy impurity measure. After all stress values are examined for reduction in impurity, the optimal stress value associated with the largest reduction in impurity is identified. That reduction becomes the Δ (i.e., reduction in impurity) for the variable of stress.

After all independent variables are considered, the independent variable with the largest reduction in impurity is selected to partition the root node into two child nodes. The CART analysis then moves on to each of the child nodes. For the left child node, for example, the same procedure can be applied to partitioning this node into two child nodes. In this way, the

CART tree keeps growing new branches, each guided by the reduction in a certain impurity measure.[2]

It is easy to sense the problem associated with the impurity measures. Impurity becomes smaller for certain when a tree grows larger. In theory, each and every tree can have a zero impurity if the tree keeps growing to yield an enormous number of terminal nodes with a single case in each terminal group (i.e., the number of terminal nodes in the tree is equal to the number of cases in the sample). Mathematically, increasing the depth or size of the tree is monotonically related to decreasing the value or degree of the impurity at the terminal nodes.

The challenge is to employ impurity to grow a tree while preventing the tree from growing too large. Breiman et al. (1984) proposed the cost-complexity measure to achieve this goal. The basic idea is to attach a penalty to the attempt to grow a large tree to reduce impurity. The larger the tree, the higher the penalty. This can be observed easily from the mathematical definition of the cost-complexity measure

$$R_{\alpha}(T) = R(T) + \alpha|T|$$

where $R(T)$ is the risk measure (the misclassification rate) of the branch or tree T, α is the nonnegative penalty coefficient, and $|T|$ is the number of terminal nodes in the branch or tree. As can be seen, large trees increase the cost-complexity measure because they produce large $\alpha|T|$.

The cost-complexity measure may guide the growth of a CART tree in a simple way or in a complex way. With $\alpha|T|$, one can think of α as the complexity cost for each terminal node. In this sense, given a desired value, the simple way to improve the cost-complexity measure is to control the number of terminal nodes in the tree. The complex or more scientific way is to search for a tree that minimizes the cost-complexity measure, $R_{\alpha}(T)$. This can be done because there is a finite number of trees between the "dead tree" (i.e., only root node without any branch) and the "mega tree" where each case is a terminal node. Of course, such a search is theoretically flawless but computationally intensive (see Breiman et al. [1984] for a possible solution).

Stopping the CART Tree

A closely related issue to the above discussion is when one should stop partitioning (i.e., terminate the growth of a CART tree). A rule (often called the stopping rule) is used to stop the partitioning process. Caution is needed when setting the stopping rule. If the rule stops the partitioning too soon, the resulting tree is likely to be too small to reflect the true structure of the data. In other words, the error in the structure of the tree tends to be large, which compromises the function of the tree. If the rule stops the partitioning too late, the resulting tree is likely to be too large to be either stable or meaningful (e.g., having few cases in terminal nodes). In other words, the tree becomes practically useless even though the error in the structure of the tree tends to be small.

There are several different ways to set the stopping rule. Traditionally, one adopts the notion of hypothesis testing to decide when to stop the tree (see Duda et al., 2001). The idea is to see whether an independent variable can perform a partition (a variable-based partition) that is statistically significantly different from a random partition. Consider a simplified case in which the dependent variable has two categories (c_1 and c_2) and a parent node has n cases (n_1 on c_1 and n_2 on c_2). If P represents the proportion of cases that an independent variable descends into the left child node, then $(1 - P)$ represents the proportion of cases that the independent variable descends into the right child node. In terms of the number of cases, the left child node receives Pn cases, and the right child node receives $(1 - P)n$ cases. Under the null hypothesis about this P, a random partition descends Pn_1 cases from c_1 and Pn_2 cases from c_2 to the left child node and the rest of the cases to the right child node. Statisticians use a chi-square (χ^2) statistic to measure the degree of deviation in the number of cases between the variable-based partition and the (weighted) random partition

$$\chi^2 = \frac{(n_{1L} - Pn_1)^2}{Pn_1} + \frac{(n_{2L} - Pn_2)^2}{Pn_2}.$$

where, under the variable-based partition, n_{1L} represents the number of cases in the left child node coming from category c_1 and n_{2L} represents the number of cases in the left child node coming from category c_2. As can be seen, the chi-square statistic increases if the variable-based partition produces an increasingly different distribution from the random partition. Recall that a critical value based on an appropriate level of significance (e.g., $\alpha = 0.05$) and degrees of freedom df are needed in a chi-square test. In this case, $df = 1$ because under a given probability P and a sample size of the parent node n, one needs only n_{1L} to figure out n_{1R}, n_{2L}, and n_{2R}. Comparing with the critical value, one either accepts the null hypothesis if the chi-square statistic is smaller (indicating a poor variable-based partition) or rejects the null hypothesis if the chi-square statistic is larger (indicating a good variable-based partition). If no independent variable can reject the null hypothesis, one stops partitioning.

Consider data from Table 2.2. Given that $n_1 = 1{,}257$, $n_2 = 1{,}380$, $n_{1L} = 503$, $n_{2L} = 387$, and P is calculated as 0.34 (i.e., 890 ÷ 2,637),

$$\chi^2 = \frac{(503 - 0.34 \times 1{,}257)^2}{0.34 \times 1{,}257} + \frac{(387 - 0.34 \times 1{,}380)^2}{0.34 \times 1{,}380} = 27.78.$$

When $\alpha = 0.05$ and $df = 1$, the critical value is 3.84. Because $\chi^2 = 27.78 > 3.84$, the variable partition (based on stress ≤ 2) is statistically significantly different from the random partition. As a matter of fact, this partition at the stress value of 2 (on a scale of 1–5) produces a more statistically significant chi-square result than partitions at any other stress values.

Another traditional approach adopts validation to decide when to stop the tree (Breiman et al., 1984). The idea is to use a subset of the data to grow the tree and use the rest of the data to validate the tree. For example, according to the conventional divide of data for the running and validation sets, one may run a CART analysis on 90% of the data and reserve the remaining 10% for the purpose of validation. The CART tree stops growing or partitioning when the error from the validation data reaches its minimum. As discussed previously, the larger the tree, the smaller the error in the structure of the tree (see the monotonic decrease in error when running or developing a tree in Figure 2.3). This general (error) trend reflects

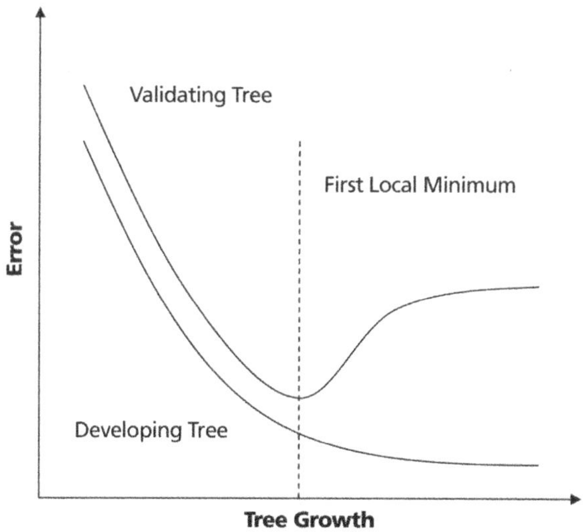

Figure 2.3 The relationships between error in tree structure and extent to which tree is developed for the running and validation data. The line representing the first local minimum indicates where one should stop growing (partitioning) tree.

also in the validation set with a monotonic decrease in the validation error until overfitting occurs in the running set. Because of overfitting, the validation error bounds back as shown in Figure 2.3. One should stop the tree when the validation error reaches its first minimum.

The smoking behavior data are rerun to illustrate the validation approach. The whole sample of 11,256 students is randomly split into a running sample (10,120 students or 90% of the whole sample) and a validation sample (1,136 students or 10% of the whole sample). Figures 2.4 and 2.5 show the results of the CART analyses. The relative risk (RR) as discussed in Zhang and Singer (1999) can be borrowed to describe the consistency between the two CART trees. At each partition, RR is defined as the percent of smokers in the left child node in ratio to the percent of smokers in the right child node, measuring the RR of smoking based on a particular independent variable. For example, using the running sample, RR = (1,102/4,050)/(3,506/6,070) = 0.47 for the partition at age = 171.5 (months; see Table 2.3).

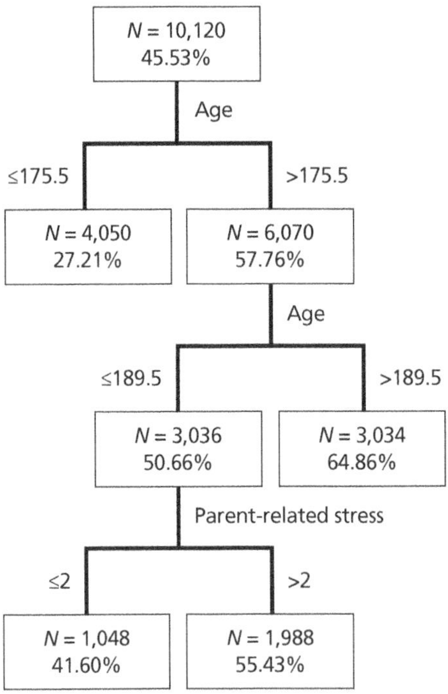

Figure 2.4 CART tree of smoking in relation to stress and background based on the running sample. In each node, the top number indicates the number of students and the bottom number indicates the proportion of smoking students (or probability of smoking).

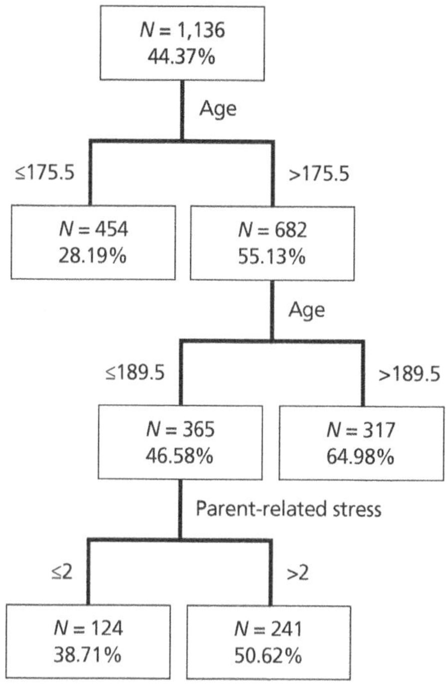

Figure 2.5 CART tree of smoking in relation to stress and background based on the validation sample. In each node, the top number indicates the number of students and the bottom number indicates the proportion of smoking students (or probability of smoking).

Overall, the RR measures in the table indicate that the two CART trees are fairly consistent in channeling students into different terminal nodes.[3] Therefore, the CART tree based on the running sample shows credible results upon validation.

In general, when using validation as the stopping rule in a CART analysis, the majority of the data is used to grow the tree (see the conventional divide above). With a large sample size, one can increase the proportion of data assigned to the validation set. With a small sample size, one often employs the cross-validation approach (called m-fold cross-validation; Breiman et al., 1984). In such an approach, one creates m (conventionally, $m = 10$) mutually exclusive subsets of the data with an equal sample size of n/m (n is the total number of cases in the root node). The CART tree is then grown m times, following the validation procedure as discussed above. In each of these m times, one leaves out one subset to function as the validation set and grows the tree on the rest of the subsets. The average of the

TABLE 2.3 Partitions of Parent Nodes Into Child Nodes Based on Smoking Data

	Non-Smoking	Smoking	Row Total	Relative Risk
Partition at Age = 171.5				
Left Child Node (R)	2,948	1,102	4,050	0.47
Right Child Node (R)	2,564	3,506	6,070	
Left Child Node (V)	326	128	454	0.51
Right Child Node (V)	306	376	682	
Partition at Age = 189.5				
Left Child Node (R)	1,498	1,538	3,036	0.78
Right Child Node (R)	1,066	1,968	3,034	
Left Child Node (V)	195	170	365	0.72
Right Child Node (V)	111	206	317	
Partition at Stress = 2				
Left Child Node (R)	612	436	1,048	0.75
Right Child Node (R)	886	1,102	1,988	
Left Child Node (V)	76	48	124	0.76
Right Child Node (V)	119	122	241	

Note: R = running sample. V = validation sample.

m classification errors is used as the validation measure which is attached to the CART tree generated from the entire (n) cases to indicate its potential performance.

Table 2.4 presents a summary of misclassification regarding the CART tree reported in Figure 2.1. The error in classification is calculated as the percentage of misclassified cases (out of the total cases). Recall that this error is actually the risk measure $R(T)$ in the cost-complexity measure. $R(T) = (1{,}944 + 1{,}935)/11{,}256 = 0.3446$ from Table 2.4. In the current example of cross-validation, ten mutually exclusive subsets are created from the whole sample of 11,256 students, with an equal sample size of 1,125 students. After the CART tree is grown ten times (based on the ten subsets), the average of the ten $R(T)$ values turns out to be 0.3486. Therefore, the result

TABLE 2.4 Results of Misclassification Based on Smoking Data

	Observed		
Predicted	**Non-Smoking**	**Smoking**	**Total**
Non-Smoking	4,209	1,944	6,153
Smoking	1,935	3,168	5,103
Total	6,144	5,112	11,256

of cross-validation is quite satisfactory, and one's confidence increases in presenting the CART tree in Figure 2.1 as the result of the analysis.

The chi-square statistic is simple but tends to be conservative. When many chi-square tests are performed on a CART tree, Type I error can be inflated. The Bonferroni method is often used to adjust the level of significance. The validation approach is somewhat superior. After all, most statisticians advocate validation in all statistical analyses. In a CART analysis, the need for validation is evident in the light of the fact that if allowed to grow freely, the tree is able to meet any criterion of accuracy (see discussion earlier). However, this accuracy is achieved in specification to a particular dataset—the tree is very likely unable to generalize to other datasets. This is exactly the logic of the validation approach desirable in a CART analysis. This approach is simple enough and can be used in the case of small samples. When using cross-validation, one needs to be aware that each case is used repeatedly ($m - 1$ times) to generate different trees. As a result, the trees are not from an independent sample, which to some extent compromises the validation summaries.

The most popular stopping rule measures the reduction in impurity.[4] Given the detailed discussion earlier on impurity measures, this approach is not strange at all. The idea is to decide on a small value as standard or threshold and compare the reduction in impurity with this standard. The CART tree keeps growing as long as there exists an independent variable that is able to reduce more impurity than the standard. When all independent variables fail to do so, one stops partitioning. The major drawback of the impurity approach lays in the difficulty in knowing how small a value is an appropriate standard. After all, impurity is a very abstract concept. Operationally, it is easier and makes more sense to work with the number of cases in a terminal node rather than the magnitude of the reduction in impurity. They are the two sides of the same "coin" of growing a tree. Recall that the bigger the tree, the purer the tree. The picture is complete when one adds the obvious, the smaller the (terminal) node. In other words, when a tree keeps splitting or growing, the reduction in impurity becomes smaller and smaller and the number of cases in the (terminal) node becomes fewer and fewer. In practice, one may stop partitioning when a node contains cases either fewer than a number-based standard (e.g., 50 cases) or smaller than a percentage-based standard (e.g., 5% of the total cases in the sample). Although the standard setting tends to be arbitrary, when properly set, the impurity approach can be an effective and efficient way to stop the tree.

Besides these statistical techniques, there are a couple of practical strategies to help one decide when to stop a tree. A group of 50 cases is often used as the minimum size of the terminal node to balance between trees too small and too large, together with the analytic strategy to limit the tree growth to a small number of levels.[5] These common analytic practices were adopted in the example in this chapter where the minimum size of any terminal node was set as 50 and the CART tree was allowed to grow three levels.

Pruning the CART Tree

"Occasionally, stopped splitting [partitioning] suffers from the lack of sufficient look ahead, a phenomenon called the horizon effect" (Duda et al., 2001, p. 402). That is, the decision to either continue or terminate partitioning at a node is made without any knowledge about its potential child nodes at subsequent levels. There is the possibility of declaring a node as a terminal one so prematurely as to sacrifice some beneficial partitions at subsequent levels. As expressed in Duda et al. (2001), "the stopped splitting biases the learning [partitioning] algorithm toward trees in which the greatest impurity reduction is near the root node" (p. 402).

The strategy to avoid this bias is straightforward—letting a CART tree grow fully until the minimum impurity standard is met everywhere in the tree.[6] One then examines all pairs of child nodes descending from the same parent nodes one level above. Any pair whose elimination leads only to a small increase in impurity is trimmed away and their parent node becomes a tentative terminal node (tentative in the sense that this node may be trimmed away with its neighboring or sibling node descending from the same parent node one level above). This method is called pruning the CART tree, the principal alternative approach to stopping the CART tree.[7] Obviously, this merging of two child nodes into a parent node (pruning) is the opposite of splitting a parent node into two child nodes (partitioning). Because, in a large CART tree, trivial reduction in impurity always occurs at the very top of the tree (away from the root node), pruning always starts from the very top of the tree working its way toward the root node. Not only does pruning avoid the horizon effect but also it utilizes all available data.[8] Duda et al. (2001) recommend that if possible pruning the tree be preferred over stopping the tree. The major drawback of pruning is that it is computationally intensive.

The CART tree in Figure 2.1 can be considered a pruned one, and it is interesting to show a CART tree before the pruning is done (see Figure 2.6). Compared with the tree in Figure 2.1, this tree contains one more branch. The parent node includes 3,356 students who are older than 189.5 months, and parent related stress partitions this node into two child nodes at the stress value of 2. For the parent node, the Gini measure is

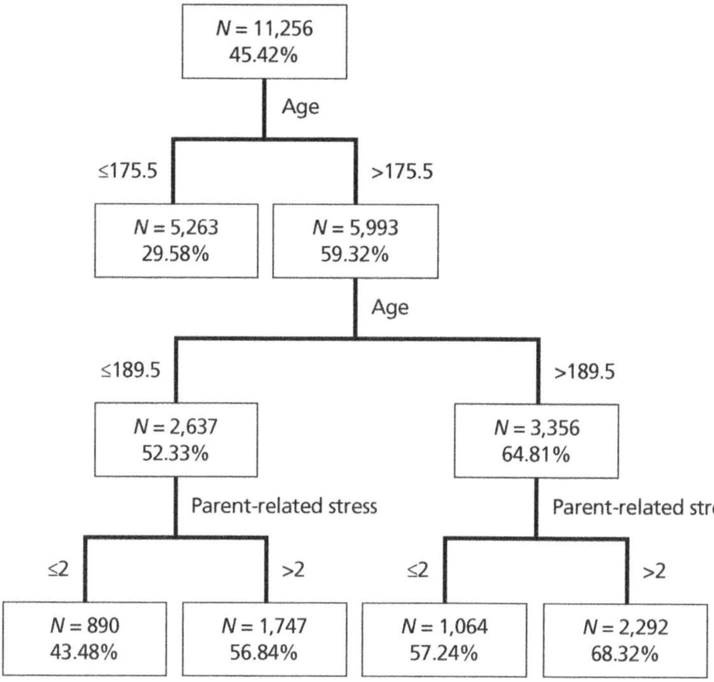

Figure 2.6 CART tree of smoking in relation to stress and background before pruning. In each node, the top number indicates the number of students and the bottom number indicates the proportion of smoking students (or probability of smoking).

$$i(\tau) = 1 - (0.3519^2 + 0.6481^2) = 0.4561.$$

Meanwhile, Gini measures for the left and right child nodes are

$$i(\tau_L) = 1 - (0.4276^2 + 0.5724^2) = 0.4895$$

$$i(\tau_R) = 1 - (0.3168^2 + 0.6832^2) = 0.4329.$$

The reduction in impurity associated with partitioning the parent node into the child nodes is

$$\Delta = 0.4561 - 0.4895 \times 0.32 - 0.4329 \times 0.68 = 0.0051.$$

Because this reduction is much smaller than the reduction associated with the neighboring partition at the same level

$$\Delta = 0.4989 - 0.4915 \times 0.34 + 0.4906 \times 0.66 = 0.0080,$$

the associated branch is trimmed away. As a matter of fact, all other partitions in Figure 2.5 have much larger reductions in impurity than this trimmed partition.

Pruning also paves the way for a successful use of the cost-complexity measure discussed earlier. Studies on the use of the cost-complexity measure show some disappointment. The major problem is that once the cost-complexity measure is directly used as a tree growth criterion, the tree tends to become unstable (i.e., poor cross-validation properties). Instead of abandoning the cost-complexity measure, Breiman et al. (1984) sought improvement by paying close attention to how it is applied to the tree growth. Originally, one grows a tree from small to large. The cost-complexity measure works poorly in this way of growing trees. However, Breiman et al. (1984) found that if one prunes a tree from large to small, the cost-complexity measure works well. This strategy leads to another popular measure of impurity called the minimal cost-complexity pruning.

Using this criterion, one starts from a very large tree. The tree can be as large as one case per terminal node. Breiman et al. (1984) emphasize that creating larger trees before one starts pruning often results in better final tree structures. Starting from a very large tree, one prunes branches successively based on the maximum reduction in the cost-complexity measure. The final tree is selected according to the one standard error rule (i.e., 1 SE rule). In the simplest form, the risk here refers to $R(T)$, the risk measure or the misclassification rate as calculated earlier in relation to data in Table 2.4. The standard error can be calculated for this risk measure as (see Breiman et al., 1984, p. 78)

$$SE = \sqrt{\frac{R(T)(1-R(T))}{N}}$$

where N is the sample size. This standard error can then be used to select the best or right sized CART tree. The procedure is to examine a group of trees of different sizes and identify the tree with the smallest $R(T)$. The corresponding SE for this tree is then calculated as above. Finally, $R(T)$ and SE add up to form a standard. The largest tree in the group with its risk measure smaller than or equal to this standard becomes the best or right sized tree.[9]

More specifically, this is how the one standard error rule works. There are many pruned sub-trees for one to choose from. The sub-tree with its risk measure within one standard error of the minimum risk measure encountered in growing the tree is selected. In cases where the risk measures of several sub-trees all meet the one standard error rule, the sub-tree with the simplest tree structure (with the smallest number of nodes) is considered the best choice.

In sum, pruning is a key component of the CART technique. Specifically, adequate tree growth followed by careful tree pruning is the very essence of a CART analysis. How large can a tree be considered a large tree? Some statisticians suggest as few as five cases (or even fewer) per terminal node. Computing intensity is the main concern in such situations. Pruning can be performed on the basis of the cost-complexity measure (minimal cost-complexity pruning), and the final tree can be selected on the basis of the one standard error rule.

Notes

1. Any tree with any number of partitions at different nodes can always be represented by a functionally equivalent binary tree. Appendix A gives a very simple illustration. Although the tree at the top part of the graph does not perform binary partition, that particular partition can be functionally equivalently represented by the tree at the bottom part of the graph that performs only binary partitions.
2. There is a caution here. Δ is a local measure that ensures the best reduction in impurity. In other words, it is specific to a local branch with one parent node descending two child nodes. Δ does not ensure that the whole (final) tree would reach optimal reduction in impurity.
3. A close match in RR between the two CART trees in Table 2.3 is a positive signal of validation. Using the notion of Figure 2.3, one may consider the close match in RR as an indication that the two CART trees are in the left vicinity of the first local minimum. RR values that are far apart indicate that the tree growth has ventured pass the first local minimum (i.e., over-fitting has occurred).
4. This statement is based on the availability of the reduction in impurity as the stopping rule in various CART software packages.
5. One may have a good reason to consider the practice that puts a limit on the number of levels in a CART tree as yet another standard to stop the tree growth, because the number of levels in a tree is not as closely associated with the reduction in impurity as the number of cases in a terminal node.
6. Any standard that specifies the minimum impurity discussed earlier can be employed here. Again, from a practical perspective, the minimum size for a node and the maximum depth (levels) of a tree are operationally simpler to specify.
7. Stopping the tree and pruning the tree are an analogy to the forward selection and backward elimination approaches in multiple regression analysis. The forward selection starts with no variables in the model and add one variable at a time, whereas the backward elimination starts with all variables in the model and delete one variable at a time. The forward selection and backward elimination approaches usually result in a very similar regression model. However,

serious differences may occur between stopping the tree and pruning the tree in CART analysis.

8. One may consider the approach of stopping a CART tree as not using all data when validation takes part in stopping the tree because part of the data is reserved for validation. The approach of pruning a CART tree uses all data because validation is usually not considered in such an approach—not necessary. In addition, a large tree often involves more independent variables than a small tree, resulting in more available data to take part in CART analysis.
9. One often asks why not just select the tree with the smallest risk measure as the final tree since it is the most accurate tree. The answer is simply that the tree may be too large because large trees tend to be more accurate in predicting cases. So the idea is to look for a simpler tree with a similar risk measure (i.e., within one standard error of the smallest risk measure).

3

Basic Techniques of CART

As pointed out earlier, one uses classification trees (CT) to grow trees in which the dependent variable is categorical, whereas one uses regression trees (RT) to grow trees in which the dependent variable is continuous. This chapter discusses the distinguishing statistical procedures of CT and RT. To emphasize that these statistical techniques are not absolutely exclusive, CART is used as a general expression where there is no need to distinguish between CT and RT.

Statistical Techniques of Classification Trees

In CART, when a parent node is partitioned, the parent node branches out to produce two child nodes. When CT partitions a parent node, it compares the impurity measure of the parent node with the impurity measures of the child nodes. The independent variable that shows the largest reduction in impurity between the parent node and the child nodes is selected to partition the parent node. Therefore, CT grows a tree using reduction in

Using Classification and Regression Trees, pages 39–57

impurity as the statistical criterion, and it performs successive binary partitions level by level.

CT allows an independent variable to appear more than once in any tree branch to capture complex relationships between the dependent variable and this independent variable, although CT performs binary partitions only. CT can handle continuous, ordinal, and nominal independent variables. CT orders values on a continuous independent variable and then examines binary partitions at all possible value points. For example, suppose there are five different values (14, 15, 16, 17, and 18) in the continuous independent variable of age, CT then examines binary partitions at four possible value points: 14 (i.e., 14 vs. 15–18), 15 (i.e., 14–15 vs. 16–18), 16 (i.e., 14–16 vs. 17–18), and 17 (i.e., 14–17 vs. 18). The partition that yields the largest reduction in impurity is then selected.

For a continuous independent variable, sometimes all values in a numerical range (a, b) result in the same maximum reduction in impurity. There are different ways to decide a cut-off point. One can use either a simple average (i.e., the midpoint), $(a + b)/2$, or a weighted average, $aP + b(1 - P)$, where P is some sort of weight based on certain preference. In the previous case of age (14, 18), the midpoint is $(14 + 18) \div 2 = 16$ (i.e., 14–16 vs. 17–18). If one prefers to partition on the younger side with, say, $P = 0.70$, the cut-off point is $14 \times 0.70 + 18 \times (1 - 0.70) = 15.20$ (i.e., 14–15 vs. 16–18).

Another reasonable strategy is to select a particular point from that numerical range that enhances certain theoretical or practical aspects related to the research questions. Note that one can also use this strategy to decide the cut-off point when a similar situation occurs with an ordinal or nominal (categorical) independent variable.

CT handles ordinal independent variables in the same way as it handles continuous independent variables. For a nominal independent variable, CT examines in an exhaustive manner all possible partitions to locate the category that maximizes reduction in impurity. In other words, CT tests all possible binary partitions of categories. For example, suppose there are three categories (White, Black, and Hispanic) in the nominal variable of race-ethnicity, then CT examines all possible two-group partitions: (White vs. Black and Hispanic), (White and Black vs. Hispanic), and (White and Hispanic vs. Black). The partition that yields the largest reduction in impurity is then selected. When working with a parent node, CT examines the reduction in impurity for each and every independent variable and selects the one that yields the largest reduction in impurity to partition the parent node.

Using Costs and Priors

Misclassification always occurs in CART as in any other classification techniques such as logistic regression, and the cost (consequence) for misclassification may not always be the same. One may not notice that the discussion in the previous chapter related to misclassification (e.g., the risk measure) is based on the assumption of equal costs for misclassification. That is, the error that misclassifies a smoker as a nonsmoker is as serious as the error that misclassifies a nonsmoker as a smoker. It is desirable in many research situations, however, to take into account misclassification costs in a CART analysis. The purpose is to correct the tree growth so as to minimize the error on the overall classification or to bring the general cost under control. Mathematically, this desire can be easily achieved by incorporating coefficients (weights) of misclassification costs into a certain impurity measure.

Take the Gini measure as an example. With some algebraic manipulation, one can rewrite the Gini measure as

$$i(\tau) = 1 - \sum P(c_j)^2 = \sum P(c_i)P(c_j)$$

where $i \neq j$. Recall that $P(c_j)$ represents the percentage of the cases falling into the category c_j (of the dependent variable) in node τ. So, the Gini measure can be alternatively described as the sum of the products of percentages of cases falling into two different categories of the dependent variable. Now, using $C(i|j)$ to denote the misclassification cost of wrongly classifying a case in category c_j into category c_i, the Gini measure can be modified as

$$i(\tau) = \sum C(i|j)P(c_i)P(c_j).$$

Obviously, each pair of different categories is given a cost (weight). Note that $C(i|j)$ may not be equal to $C(j|i)$. So, the modified Gini measure is the weighted sum of the products of percentages of cases falling into two different and ordered categories of the dependent variable. Incorporating misclassification costs into a CART analysis can affect both the tree structure (i.e., the way that the tree is partitioned) and the case assignment (i.e., the way that cases are channeled into terminal nodes). The risk measure for the new tree can also change.

The CART tree in Figure 2.1 is developed without consideration of misclassification costs. The case of equal misclassification costs is illustrated in Table 3.1 in which the number zero indicates a correct classification

TABLE 3.1 Incorporating Misclassification Costs Into CT Analysis

Equal Misclassification Costs		
	Observed	
Predicted	**Non-Smoking**	**Smoking**
Non-Smoking	0	1
Smoking	1	0
Unequal Misclassification Costs		
	Observed	
Predicted	**Non-Smoking**	**Smoking**
Non-Smoking	0	2
Smoking	1	0

whereas the number one indicates a misclassification cost. This table also shows a case where unequal misclassification costs are introduced. Specifically, misclassifying a smoker as a nonsmoker costs twice as much as misclassifying a nonsmoker as a smoker. As a result, one wants to misclassify fewer smokers. Using c_1 as the category of nonsmokers and c_2 as the category of smokers, this specification means that $C(2|1)=1$ and $C(1|2)=2$. Continuing to work with data in Table 2.2, the (modified) Gini measure for the left child node is

$$\begin{aligned} i(\tau_L) &= C(1|2)P(c_1)P(c_2)+C(2|1)P(c_2)P(c_1) \\ &= 1\times 0.5652\times 0.4348+2\times 0.4348\times 0.5652 \\ &= 0.7372. \end{aligned}$$

Similarly, $i(\tau_R)=0.7435$ and $i(\tau)=0.7483$. The reduction in impurity is now

$$\Delta = 0.7483-0.7372\times 0.34-0.7435\times 0.66 = 0.0069.$$

Note that this reduction is quite a drop from the reduction in impurity of 0.0080 obtained without specifying unequal misclassification costs.

Figure 3.1 presents the results of a new CART analysis of the smoking data incorporating misclassification costs as specified above. The tree structure is quite different from that in Figure 2.1. With unequal misclassification costs, the attention is now shifted to younger students (a new partition at the age value of 156.5 months). When misclassifying a smoker as a nonsmoker costs twice as much as misclassifying a nonsmoker as a smoker,

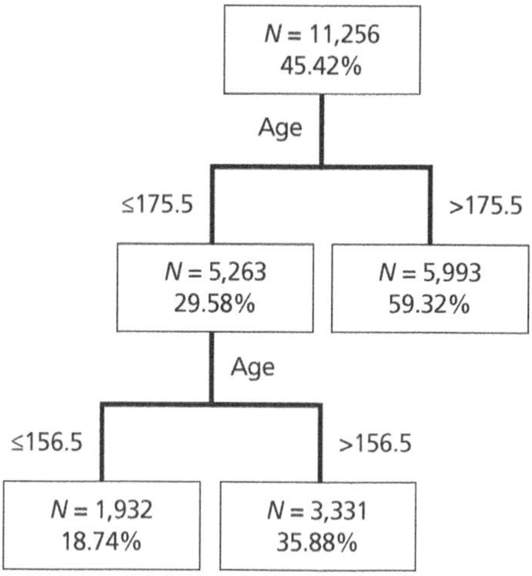

Figure 3.1 CART tree of smoking in relation to stress and background with costs of misclassification specified. In each node, the top number indicates the number of students and the bottom number indicates the proportion of smoking students (or probability of smoking).

effects associated with parent related stress disappear. On the other hand, the critical factor for starting smoking, age, becomes even more critical. The three terminal nodes are all about age effects, indicating that older students are increasingly likely to smoke (about 19%, 36%, and 59% respectively across age groups).

Table 3.2 presents misclassification data regarding the new CART tree. As can be seen, there are a lot more predicted smokers in Table 3.2 (the number is 9,324) than in Table 2.4 (the number is 5,103). This change reflects the higher cost of misclassifying a smoker as a nonsmoker. Specifying equal costs tends to balance the misclassification rates as evidenced in Table 2.4 (1,944 versus 1,935). Increasing the cost for misclassifying smokers tends to decrease the misclassification rate on smokers but meanwhile

TABLE 3.2 Misclassification Based on Smoking Data After Specifications of Costs

	Observed		
Predicted	**Non-Smoking**	**Smoking**	**Total**
Non-Smoking	1,570	362	1,932
Smoking	4,574	4,750	9,324
Total	6,144	5,112	11,256

increase the misclassification rate on nonsmokers as evidenced in Table 3.2 (362 versus 4,574). Finally, the risk measure $R(T)$ is calculated as

$$R(T) = (2 \times 362 + 1 \times 4{,}574)/11{,}256 = 0.4707.$$

Priors are another way to take into consideration misclassification costs in a CART analysis. Priors refer to prior knowledge, experiences, or expectations about a certain population, and "intelligent selection and adjustment of them can assist in constructing a desirable classification tree" (Breiman, Friedman, Olshen, & Stone, 1984, p. 112). Three things are essential to understand how priors work. First, priors affect misclassification costs. Suppose the dependent variable has two categories. If prior knowledge indicates that the probability of one category c_1 occurring is twice that of the other category c_2 occurring, then it naturally costs twice as much to misclassify a case from c_1 to c_2.[1] In other words, if a misclassification from c_2 to c_1 counts as one error, then a misclassification from c_1 to c_2 must count as two errors. This logic links priors with costs—specifying a larger prior probability for a certain category increases the misclassification cost for that category.

Second, as probabilities, priors for c_1 and c_2 must add up to one. Going back to the smoking data, there are reasons to believe that the majority of students do not smoke (e.g., campaign against smoking). Majority may be conservatively defined as 3/5. Therefore, the prior for nonsmoking is 0.60, whereas the prior for smoking is 0.40 (this implies that misclassifying a nonsmoker is 1.50 times more expensive than misclassifying a smoker). When priors are incorporated into a CART analysis, the tree partition and the case assignment can be changed. Indeed, a reanalysis of the smoking data with the priors specified above produces a different tree structure (see Figure 3.2).

Compared with Figure 2.1, with other parts of the tree remaining intact, the tree branch associated with parent related stress is trimmed away because of its trivial reduction in impurity after priors are taken into analysis. When prior knowledge is incorporated into the CT analysis, age becomes the only critical factor in determining smoking among students. Table 3.3 presents misclassification information for the new tree. There are now more predicted nonsmokers in Table 3.3 where the number is 7,900 than in Table 2.4 where the number is 6,153 because of the specification that there are more nonsmokers than smokers in the population. This result makes sense because priors dictate misclassification costs. The risk measure $R(T)$ is calculated as

$$R(T) = (0.60 \times 2{,}937 + 0.40 \times 1{,}181)/11{,}256 = 0.1985.$$

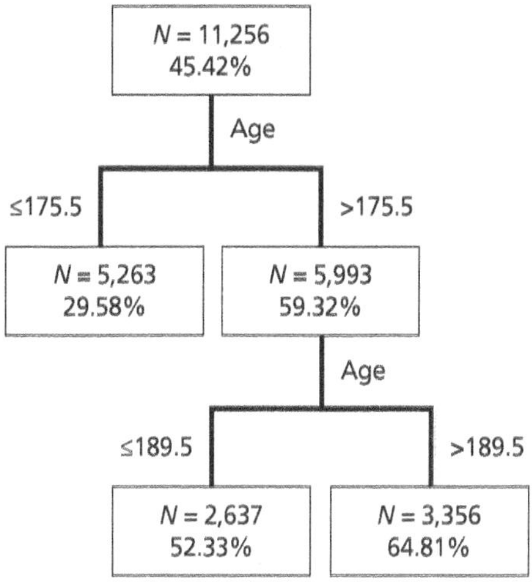

Figure 3.2 CART tree of smoking in relation to stress and background with priors specified. In each node, the top number indicates the number of students and the bottom number indicates the proportion of smoking students (or probability of smoking).

TABLE 3.3 Misclassification Based on Smoking Data After Specifications of Priors

	Observed		
Predicted	**Non-Smoking**	**Smoking**	**Total**
Non-Smoking	4,936	2,937	7,900
Smoking	1,181	2,175	3,356
Total	6,144	5,112	11,256

The third thing essential to understand how priors work is that the misclassification cost is the same no matter into which category a case is misclassified from its correct category. Consider a simple case in which a dependent variable has three categories c_1, c_2, and c_3 with three priors or prior probabilities. The cost is the same between misclassifying c_1 into c_2 and misclassifying c_1 into c_3, even though c_2 and c_3 may have different priors (indicating different misclassification costs).[2]

If a dependent variable has two categories, using either misclassification costs or prior probabilities produces equivalent analytical results. Recall that the priors specified above imply that the cost to misclassify a nonsmoker is 1.50 times as much as that to misclassify a smoker. Using this information as costs to run a new analysis would produce the same analytical results. When a dependent variable has more than two categories, using misclassification costs produces different analytical results from using prior probabilities.

In practice, for the purpose of control for the tree growth, misclassification costs are based more on preferences, whereas prior probabilities are based more on facts. For example, one may use costs to purposefully minimize misclassification on a certain category, whereas one may use priors to objectively adjust the under-sampling of certain categories if a sample is not fully representative of a population.[3]

If a departure from the correct category costs the same no matter where a misclassified case falls, prior probabilities are a good choice. Otherwise, one can use misclassification costs to specify the differences in cost between misclassifying c_i into c_j $(i \neq j)$ and misclassifying c_i into c_k $(i \neq k)$. Costs and priors can also be employed jointly to, for example, both correct undersampling biases and control misclassification rates.

Costs and priors give the risk measure $R(T)$ different meanings. Table 3.4 attempts to help one correctly understand and interpret the risk measures under different combinations of costs and priors. As far as priors are concerned, they are given different values in the above example (i.e., specified). Priors can also be specified as equal for all categories of the dependent variable. There is another way of handling priors. The default of many CART programs assumes that the sample distribution (in terms of the proportions of cases falling into each category of the dependent variable) reflects the population distribution. These proportions are called empirical priors. As long as priors are specified (i.e., not empirical), the risk measure is for the population that matches the set of priors for the analysis but not for the sample with which one is working. For example, in Table 3.3, the risk measure for the sample would be $(2{,}937 + 1{,}181)/11{,}256 = 0.3658$ (different from the $R(T)$ value calculated earlier). Although using costs produces equivalent analytical results to using priors when a dichotomous dependent variable is used, the risk measure is different both conceptually and numerically.

TABLE 3.4 Meanings of the Risk Measure *R*(*T*) Under Different Combinations of Costs and Priors

Costs	Priors	Meanings of the Risk Measure $R(T)$
Equal to one	Empirical	Expected Error Rate
Equal to one	Specified	Expected Error Rate for a Population Matching the Priors
Unequal	Empirical	Expected Cost of Errors
Unequal	Specified	Expected Cost of Errors for a Population Matching the Priors

Source: SPSS (1999).

When costs are equal (to 1), the term, rate, in Table 3.4 can be simply interpreted as a probability of making errors. When costs are unequal, $R(T)$ becomes the cost of making errors rather than the probability of making errors. In this case, it is not uncommon to have the $R(T)$ value greater than 1 (because $R(T)$ is no longer a probability). Finally, there is a caution for using costs and priors. Because costs and priors have certain undesirable properties (see Breiman et al., 1984, Chapter 4), the motivation to employ costs and priors needs to be justified carefully.

Statistical Techniques of Regression Trees

Like CT, RT performs binary partitions of nodes successively based on a statistical criterion (the Gini measure in the case of CT), and when working with a parent node, the independent variable that yields the largest improvement in the criterion (i.e., the largest reduction in impurity) is selected to partition the parent node into child nodes. Because the dependent variable is now continuous, the definition of impurity under CT that relates to the categories of the dependent variable is no longer appropriate. Instead, the within-node variance naturally becomes the focus of the criterion. The idea is to minimize the within-node variance so as to produce nodes that are as homogeneous as possible on the dependent variable. A node is called pure when all cases in that node share the same value on the dependent variable, whereas a node is called impure when cases in that node show diverse values on the dependent variable. Therefore, the within-node variance becomes the impurity measure for RT.

The within-node variance measures the degree to which responses from cases (1, 2, 3,... n) within a node (τ) spread out along the dependent variable. To calculate, one uses the sum of squared deviations (also called the sum of squares)

$$i(\tau) = \sum (y_i - y.)^2$$

where y_i is the value on the dependent variable for case i ($i = 1, 2, 3, \ldots n$) and $y.$ is the (node) mean of the dependent variable. When RT partitions a parent node, it compares the impurity measure of the parent node with the impurity measures of the child nodes. The independent variable that shows the largest reduction in impurity between the parent node and the child nodes is selected to partition the parent node

$$\Delta = i(\tau) - i(\tau_L) - i(\tau_R).$$

Note that unlike CT, no weights are necessary in calculating reduction in impurity in RT (e.g., Breiman et al., 1984; Zhang & Singer, 1999). This formula also implies the way that one can calculate impurity for a branch or even a tree. When a parent node is partitioned into two child nodes, RT uses the sum of the impurity measures of those child nodes as the impurity measure for the branch.

Therefore, like CT, RT grows a tree using reduction in impurity as the criterion, performing successive binary partitions level by level. Unlike CT in which impurity ranges between 0 and 1, impurity in RT has no upper boundary. This situation requires that when one evaluates reduction in impurity, the original variance of the dependent variable (or the within-node variance of the root node) be used as the baseline or reference.

The within-node variance as a measure of impurity for the tree growth suffers the same problem as the impurity measures in CT. That is, substantial reduction in impurity can be achieved by enlarging a tree. Theoretically, each and every tree can have a zero impurity when each terminal group has only one case (the within-node variance is zero). The cost-complexity measure is used to address this problem.

Using Cost Complexity

The principle is to attach a penalty to a large tree. The definition formula is the same as that in CT

$$R_\alpha(T) = R(T) + \alpha|T|.$$

The difference is that in RT the risk measure is just the within-node variance of the tree (or the sum of impurity measures of all terminal nodes)

$$R(T) = \sum i(\tau).$$

Therefore, to improve the cost-complexity measure, one needs to reduce the risk (the within-node variance) and to keep the complexity penalty under control.

The cost-complexity measure in RT runs into the same problem as that in CT—this measure is not satisfactory as a tree growth criterion because it has the tendency to build unstable tree structures. To avoid this problem, one needs to grow a very large tree first and then use the cost-complexity measure to prune the tree—the criterion of the minimal cost-complexity pruning (see discussion in the previous chapter). Using this criterion, one starts with a very large tree, and prunes branches successively based on the maximum reduction of the cost-complexity measure. The final tree is selected based on the one standard error rule (1 SE rule)—a pruned subtree with its risk measure within one standard error of the minimum risk measure found in growing the tree is considered the best candidate for the final selection (see discussion in the previous chapter). If there are several sub-trees with their risk measures meeting the one standard error risk rule, the one with the simplest tree structure (i.e., with the smallest number of nodes) is considered the final choice.

Costs and priors are in general not a concern in RT. Although costs are still relevant in concept, costs for misclassified categories are not easily definable for a continuous dependent variable. In fact, costs are now the difference (or distance) between the observed and predicted values. As mentioned above, in RT, the risk measure addresses the cost-complexity issue and connects with the within-node variance. Therefore, the within-node variance, to some extent, captures the concept of costs. Even though priors can still be taken into account in an RT analysis, they are primarily used to match a sample distribution to a population distribution. The function of priors to influence costs, as one sees in CT, is no longer available in RT. If it is highly desirable to incorporate costs and priors into an RT analysis, one can consider categorizing a continuous dependent variable into a dichotomous dependent variable by rationalizing the cut-off point. Of course, this treatment turns an RT analysis into a CT analysis.

Like CT, RT also permits an independent variable to appear more than once in any tree branch to discover complex relationships between the dependent variable and this independent variable, although RT performs binary partitions only. The way that RT handles continuous, ordinal, and nominal independent variables is exactly the same as the way that CT handles continuous, ordinal, and nominal independent variables. The discussion so far on CT and RT clearly indicates that these two techniques differ only in specific (also minor) details.

Figure 3.3 represents another CART analysis (more precisely an RT analysis) of the relationship between smoking behaviors of young adolescents and their parent (home) and teacher (school) related stress, with a different indicator (or dependent variable) of smoking—the number of cigarettes smoked weekly. The current sample (N = 11,226) is essentially the same as the one analyzed in the previous chapter (N = 11,256), with some cases removed due to missing values on the dependent variable. Independent variables remain unchanged: parent related stress, teacher related stress, gender, age, and the number of parents. This analysis attempts to test the research hypothesis that

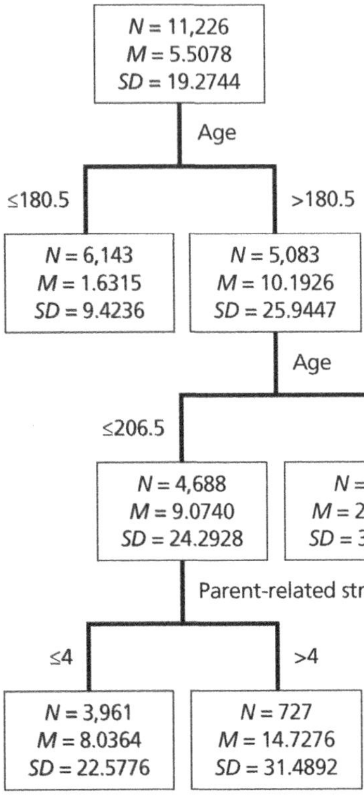

Figure 3.3 CART tree of smoking in relation to stress and background. In each node, *N* indicates the number of students, *M* indicates mean in the number of cigarettes smoked, and *SD* indicates standard deviation in the number of cigarettes smoked.

stress is related to smoking behaviors among young adolescents from a different point of view, namely how much adolescents smoke (as opposed to how likely adolescents smoke in the previous chapter).

The RT tree in Figure 3.3 indicates that age and parent related stress are the most successful independent variables to partition students who smoke a certain number of cigarettes. Age is a particularly successful independent variable in that it successfully partitions students two times in the tree. To provide further insights, the mean values of all independent variables are presented in Table 3.5 for all terminal nodes in the RT tree. This table is arranged purposefully based on the average number of cigarettes smoked from small to large.

The oldest students (395 in Node 4) in the sample smoked the largest number of cigarettes weekly (23 to 24 cigarettes), more than four times

TABLE 3.5 Stress and Background Characteristics for Each Terminal Node in the RT Tree

	Node 0	Node 1	Node 2	Node 3	Node 4
Parent Related Stress (Scale of 1–5)	2.91	2.75	2.73	5.00	3.12
Teacher Related Stress (Scale of 1–5)	2.49	2.35	2.59	3.00	2.51
Gender (Proportion of Male Students)	0.52	0.52	0.53	0.51	0.45
Age (132–241 Months)	177.12	162.84	192.76	192.35	213.96
Number of Parents (0–2)	1.79	1.79	1.81	1.81	1.73

Note: Node 0 represents the root node. Node 1 represents the terminal node at the first level. Nodes 2 and 3 represent the terminal nodes at the third level (from left to right). Node 4 represents the terminal node at the second level. Terminal nodes are arranged according to the mean number of cigarettes smoked weekly from the least to the most.

higher than the mean of the sample (5 to 6 cigarettes). The age of these 395 students averages 213.96 months. These students also have above average stress related to parents (3.12 versus 2.91) and teachers (2.51 versus 2.49). In addition, about 45% of these students are male, and the average number of parents is below the mean of the sample (1.73 versus 1.79). Overall, this group of high-risk students can be characterized as being older, having more females than males (the only such case across the four terminal nodes), being more likely to come from single-parent families, and having above average stress related to parents and teachers.

The 727 students in Node 3 demonstrate the second largest number of cigarettes smoked weekly (14 to 15 cigarettes), almost three times higher than the mean of the sample (5 to 6 cigarettes). These students are younger (all ≤ 206.5 months but > 180.5 months, mean = 192.35 months) but have highest parent related stress (mean = 5.00 in a scale of 1–5) and teacher related stress (mean = 3.00 in a scale of 1–5) in the sample. About 53% of these students are male, and they are one of the two groups of students who are less likely to come from single-parent families (mean = 1.81).

The 3,961 students in Node 2 smoked 8 to 9 cigarettes weekly, nearly 50% higher than the mean of the sample (5 to 6 cigarettes). These students are younger than those in Node 4 (all ≤ 206.5 months, mean = 192.76 months), and they have below average parent related stress (all ≤ 4 on a scale of 1–5, mean = 2.73) but above average teacher related stress (mean = 2.59). About 53% of these students are male, and they are the other group of students who are less likely to come from single-parent families (mean = 1.81).

Finally, the youngest students (6,143 in Node 1) in the sample smoked the smallest number of cigarettes weekly (1 to 2 cigarettes), three times lower than the mean of the sample (5 to 6 cigarettes). The age of these students averages 162.84 months. They have below average parent related stress (2.75 versus 2.91) and teacher related stress (2.35 versus 2.49). About 52% of these students are male, and this group of students reflects (or is representative of) the sample in terms of the number of parents (mean = 1.79).

Clearly, Nodes 3 and 4 represent students at high risk of "excessive" smoking. This high-risk group accounts for almost 10% (i.e., (727 + 395)/11,226) of the population (i.e., students in Grades 6 to 10). That is, one in ten students in Grades 6 to 10 is at risk of excessive smoking. Students in Node 2, about 35% (i.e., 3,961/11,226) of the population, are also at somewhat high risk of smoking. On the other hand, Node 1 represents students at low risk of smoking. This low-risk group accounts for about 55% (i.e., 6,143/11,226) of the population. Note that students at high risk of smoking add up to a substantial 45% of the student population in Grades 6 to 10. These students concentrate at upper junior high school and low senior high school (Grades 9 and 10) in the current sample of students in Grades 6 to 10. These grades, thus, should be the focus of smoking prevention and intervention.

Focusing on stress, one can see from Table 3.5 that parent related stress almost doubles the number of cigarettes smoked weekly. That is, students under high parent related stress smoke nearly twice as much as students under low parent related stress. Therefore, reducing parent related stress is an effective strategy to reduce the amount of smoking among students in junior high school. Mentioning junior high school is important because parent related stress distinguishes the amount of smoking not for all students attending Grades 6 to 10 but for students with an age range from 180.5 to 206.5 in months, which indicates junior high school grades. Parent related stress does not make a significant difference in the amount of smoking for students in other age ranges.

Using R-Squared

Some demonstrations are in order now to show partitions in the RT tree. Table 3.6 lists within-node variances for all (parent and child) nodes in Figure 3.3. Given information in Table 3.6, the reduction in impurity from the root node to the child nodes can be easily calculated as

$$\begin{aligned}\Delta &= i(\tau) - i(\tau_L) - i(\tau_R)\\ &= 4{,}170{,}481.01 - 545{,}524.43 - 3{,}421{,}506.87\\ &= 203{,}455.71\end{aligned}$$

TABLE 3.6 Within-Node Variances for Parent and Child Nodes

Node	First Partition	Second Partition	Third Partition
Parent Node	4,170,487.01	3,421,506.87	2,766,576.94
Left Child Node	545,524.43	2,766,576.94	2,019,111.91
Right Child Node	3,421,506.87	580,826.33	720,871.18

Note: The within-node variance is calculated as the product of squared standard deviation and sample size in each node in Figure 3.3 to get the sum of squares. The first partition occurs from the root node to the first level of the tree. The second partition occurs from the first to the second level of the tree. The third partition occurs from the second to the third level of the tree.

Note that this partition effectively polarizes high-value and low-value cases (the number of cigarettes smoked weekly in the current case) into the child nodes, although the goal of this (as well as each and every) partition is to reduce the within-node variance. Getting within-node variances from all terminal nodes, one can calculate the risk measure

$$
\begin{aligned}
R(T) &= \sum i(\tau) \\
&= 545{,}524.43 + 580{,}826.33 + 2{,}019{,}111.91 + 72{,}019{,}871.18 \\
&= 3{,}866{,}333.85.
\end{aligned}
$$

The variance deduction between the root node and the terminal nodes is then

$$4{,}170{,}487.01 - 3{,}866{,}333.85 = 304{,}153.16.$$

Similar to the R^2 in multiple regression analysis, a pseudo R^2 can be calculated in RT as

$$304{,}153.16/4{,}170{,}487.01 = 0.07.$$

Therefore, about 7% of the variance in the root node has been explained by the RT tree.[4] This amount is certainly not as large as one may want to see.

The borrowing of the concept of R^2 from multiple regression analysis provides one with a way to evaluate the tree performance, similar to the way that R^2 is used to evaluate the model performance in multiple regression analysis. The pseudo R^2 for RT is commonly discussed in the literature. With a little extension, one can actually evaluate the relative contribution of each terminal node to the tree performance

$$1 - i(\tau)/R(T).$$

The idea is to "award" a terminal node with a small variance because cases in this terminal node are more homogenous. Going back to Figure 3.3 and Table 3.6, one can identify a terminal node at the first level of the tree or the first partition—the left child node with the impurity measure of 545,524.43. The relative contribution of this terminal node is then

$$1 - (545{,}524.43/3{,}866{,}333.85) = 0.86.$$

The relative contribution can be considered a simple index ranging from 0 to 1 with a larger value indicating a more important contribution. Table 3.7 presents the indices of the relative contribution for all terminal nodes in Figure 3.3 (and Table 3.6). The left child node at the first partition and the right child node at the second partition are the real "anchors" of the tree with the largest indices (0.86 and 0.85 respectively in relative contribution). They make important contribution to the tree. It is often the case that terminal nodes closer to the root node indicate more contributions because it is easier to find a homogeneous group when the sample is large (i.e., at the beginning with the whole sample). As the partition goes on, sample size gets smaller and smaller. It becomes harder to find a homogenous group when there are a small number of cases. From this perspective, the right child node at the third partition makes important contribution to the tree with 0.81 as the index.

One may have been convinced from the current case that the within-node variance as an impurity measure can become extremely large. Recall that this impurity measure has no upper boundary. The fact that this impurity measure is a function of the (node) sample size can greatly inflate its value. Therefore, it is imperative, as mentioned earlier, to assess the reduction in impurity in reference to the within-node variance in the root node. Depending on the within-node variance in the root node, a reduction in

TABLE 3.7 Relative Contributions of Terminal Nodes to Tree Performance

Terminal Position	Variance	Relative Contribution
Left child node at the first partition	545,524.43	0.86
Right child node at the second partition	580,826.33	0.85
Left child node at the third partition	2,019,111.91	0.48
Right child node at the third partition	720,871.18	0.81

Note: $R(T)$ = 3,866,333.85

impurity of, say, 1,000 can be trivial or enormous. In the current case, when the root node is partitioned into two child nodes, the above calculation indicates that the reduction in impurity is quite marginal, about 5% of the within-node variance of the root node (i.e., 203,455.71/4,170,487.01 = 0.05).

Using Surrogates

One of the distinguished characteristics of CART is that it performs multiple single-variable partitions when attempting to partition a parent node into child nodes (and then picks up the independent variable with the largest reduction in impurity). Therefore, a record exists for all independent variables to show their performance (in partitioning the parent node). Although one independent variable is selected to partition the parent node, one may still ask whether there are other independent variables doing nearly as well as the chosen one in channeling cases from the parent node to the child nodes. Table 3.8 presents such a record taken for the partition from the root node to its child nodes based on the RT analysis in Figure 3.3.

The table lists two best independent variables that can be used to reproduce partitions performed by the chosen independent variable of age. These best candidates are rank ordered according to their association with the chosen independent variable. Note that Wilks' Lambda (λ) for contingency tables is often used to evaluate improvement in classification and can be used here as a measure of association (see Tabachnick & Fidell, 2007).[5] This measure indicates the degree to which partitions made by an independent variable match those made by the chosen independent variable. An independent variable with a high association value is a good candidate to substitute the chosen independent variable. The measure of association ranges from 0 to 1.

With such information as presented above, one can have a reasonably good idea about which independent variables are able to closely replicate (or reproduce) partitions performed by the chosen independent variable. These independent variables are called surrogates. In the current case, the best surrogate is teacher related stress, with an association value of 0.07. The reduction in impurity is 8,682.32. Surely, one wants to have a much

TABLE 3.8 Surrogates of Age for the Partition of the Root Node

Surrogate	Partition	Association	Reduction in Impurity
Teacher-Related Stress	≤2, >2	0.07	8,682.32
Parent-Related Stress	≤3, >3	0.03	27,887.83

stronger association and a much larger reduction in impurity for a surrogate. But given that some parent nodes, such as the one with students older than 180.5 months (i.e., the parent node at the first level of the tree in Figure 3.3), may have no suitable surrogates at all, teacher related stress is considered a workable substitute independent variable for the chosen independent variable in that it produces partitions similar to those performed by the chosen independent variable.

One can easily sense the advantage of surrogates when dealing with missing values on the chosen independent variable. Indeed, surrogates are used in the CART technique to assign cases with missing values on the chosen independent variable to the appropriate child node. Some software programs for CART such as the SPSS Decision Tree use surrogates as their default function to handle missing values. In the current example, when a case has a missing value on age, its assignment to one of the child nodes is determined based on its value on teacher related stress.

Surrogate analysis is also a valuable tool by itself. It shows which independent variables are associated or unassociated with the chosen independent variable (in terms of partitioning a parent node into child nodes). For example, if there exist a few quite excellent surrogates, policy implications become flexible because several different policy options (based on these surrogates) are able to achieve the same function. Consider another example. When designing instruments (e.g., questionnaires), if information on a certain item is not easily available, knowledge about its surrogates helps one find alternatives.

Note that in Table 3.8, priorities are rank ordered on the basis of association (or how closely partitions made by a surrogate match those made by the chosen independent variable). As easily seen in the table, a better surrogate (for partitioning a parent node) may not necessarily have a better reduction in impurity. Rearranging priorities on the basis of reduction in impurity in the table creates a new list, called a predictor list. This is another record that most CART software programs keep in memory. Therefore, one may ask this question: Among all available independent variables which one is both an excellent predictor and an excellent surrogate in comparison to the chosen independent variable? In most analyses, the best predictor (next to the chosen one) is indeed the best surrogate. This consistency increases one's confidence in using surrogates to inform theories, policies, or practices.

Notes

1. In a heuristic sense, because the probability of falling into category c_1 is twice that of falling into category c_2, each case that belongs to c_1 represents 2 points and each case that belongs to c_2 represents 1 point. If a case belonging to c_1 is

misclassified, it suffers a 2-point loss. If a case belonging to c_2 is misclassified, it suffers a 1-point loss. In this way, it costs twice as much to misclassify a case from c_1 to c_2.

2. One may attach some heuristic meanings to this example. Suppose the probability of falling into category c_1 quadruples that of falling into category c_2 and doubles that of falling into category c_3. Then, each case that belongs to c_1 represents 4 points, each case that belongs to c_2 represents 1 point, and each case that belongs to c_3 represents 2 point. So, misclassifying a case belonging to c_1 suffers a 4-piont loss no matter the case gets into c_2 or c_3.
3. One may rightfully argue that both costs and priors are a tactic (purposeful) use of analytical strategy. Nonetheless, on top of this, costs are often set by preferences (so as to continue the tactic orientation), whereas priors are often set by facts (so as to discontinue the tactic orientation).
4. The concept of R^2 is obviously appropriate for RT. From the previous discussion on CT, one can see that the issue of within-node variance is not relevant when a node is categorical (i.e., the dependent variable is categorical so that each node in the case of a CT tree contains information on how many cases fall into each category). However, there are other ways to adopt the concept of R^2 for CT (even though one may not consider it natural for CT). This issue is dealt with in Chapter 4.
5. For contingency tables, measures of association are usually symmetric, meaning that the value of the association remains the same no matter which variable is used as the dependent variable. If a measure of association changes depending on the selection of row or column as the dependent variable, this measure is often called asymmetric. λ can be symmetric or asymmetric. Here it takes the symmetric form.

4

Issues in CART Analysis

CART Versus Traditional Statistical Techniques

CART can be considered a heuristic tree method that unearths the relationships embedded in the data. Variables participating in CART can be either categorical or continuous. CART produces a summary tree diagram that indicates which independent variables are associated with the dependent variable and how interactions among independent variables generate groups with varying average measures on the dependent variable.

The word, regression, in CART sometimes makes one ask the question of why CART when there is multiple regression. CART advantages were discussed earlier. Apart from those general advantages (over traditional statistical techniques), Morrison (1998) presented specific advantages of CART over traditional regression techniques. CART makes it much easier than regression to gain an in-depth insight into different segments of data, handle missing data, and capture nonlinearity and interactions within data.

Traditionally, logistic regression is used to handle analyses with dichotomous dependent variables. In comparison to CT, logistic regression yields

Using Classification and Regression Trees, pages 59–74

a better performance when most independent variables are continuous and one does not expect complex interactions among those independent variables. If most independent variables are categorical or ordinal, and especially if one expects that those independent variables are related by interactions to the dependent variable, then CT yields a superior performance to logistic regression.

The ability of RT to identify nonlinear relationships and interaction effects is described as far more "native" (or superior) than multiple regression in training manuals of software packages such as SPSS. Being a non-parametric method, RT is more robust to the negative impact of outliers and abnormal distributions that is fatal to multiple regression (Breiman, Friedman,Olshen, & Stone, 1984). Therefore, RT should be seriously considered over multiple regression, especially when one expects the presence of nonlinear relationships, interaction effects, outliers, or abnormal distributions. In addition, RT has a classification or prediction function in that it channels cases into different terminal nodes with dramatically different outcomes (on the dependent variable), but these group dynamics are hard to capture with multiple regression. SPSS training manuals describe RT as more efficient computationally than many new predictive techniques such as kernel methods and robust regression, making it extremely attractive in situations where there are a large number of predictor variables.

For some, the word, classification, in CART immediately brings to mind factor analysis and cluster analysis. These traditional statistical techniques have the function of data reduction. Factor analysis detects groups (factors) among variables, while cluster analysis detects groups among either variables or cases. At first sight, CART seems to resemble factor analysis and cluster analysis in that all of these statistical techniques do classifications. There are important differences, however. CART clusters cases based on their independent variables in relation to a dependent variable. In neither factor analysis nor cluster analysis, however, is there a dependent variable. Therefore, to some extent, one can reasonably conclude that CART actually classifies relationships between the dependent variable and independent variables as they manifest differentially among groups of cases. Apart from this important distinction, there may not be overwhelming advantages of CART over cluster analysis.

Practically, here are the major differences between cluster analysis and CART. One chooses the number of groups in cluster analysis whereas one leaves the number of groups to be determined by data in CART (better control for cluster analysis versus better results for CART). Cluster analysis splits data using all variables altogether whereas CART splits data using one variable at a time (cleaner tree for CART). Cluster analysis accommodates continuous data whereas CART accommodates continuous and categorical data (more flexible for CART). Cluster analysis often works with a moderate sample for data analysis whereas CART often works with a large sample (more applicable for cluster analysis). Finally, cluster analysis requires some skills to perform analysis as one manipulates different diagnostics to obtain the best result whereas CART requires some data preparation to produce a set of clearly defined variables to generate a clean tree that is easy to interpret and understand (an issue of skills versus efforts).

Overall, CART is designed to handle far more independent variables than its traditional parametric counterparts can handle. These traditional parametric techniques demand a lot on sample size and data distribution, with the increase in the number of independent variables. CART avoids this problem successfully because it works with only one variable at a time within each node. In this way, CART can handle an enormous number of independent variables at the cost of losing some statistical coverage (e.g., lack of covariance perspective). But there is not much concern about this loss given that the model data fit is seldom satisfactory for traditional parametric techniques anyway. Some statisticians consider the presence of a single mathematical equation that clearly defines and quantifies the relationship between the dependent variable and independent variables as an advantage of the traditional parametric techniques over CART. Indeed, the tree structure in CART cannot be expressed through any mathematical equations. This "handicap," however, as argued in Chapter 1, may not necessarily be a bad thing in that many applied researchers prefer to avoid complex mathematical expressions for straightforward interpretation and understanding of analytical results.

It is always a sound statistical practice to apply different statistical techniques to examine the same data at hand. One may refer to this statistical practice as statistical triangulation. The reconciliation of differences from various statistical techniques often increases one's confidence in making a credible knowledge claim. Cluster analysis and CART, for example, are especially in such a relationship because both statistical techniques aim to segment data as a way to understand the relationships embedded in the data.

Formulating Research Questions

The overall principle is to ask research questions that take full advantage of the CART capabilities. Given the main structure of any CART analysis (i.e., the CART tree), one may approach this issue from the within node and between node perspectives. From the within node perspective, because

CART creates homogenous groups (i.e., terminal nodes), these "like-minded" cases (e.g., individuals) within each node share common behaviors. Research questions may pertain to the characteristics of the high outcome and low outcome groups. For example, what are individual and family characteristics of adolescents who are most and least likely to smoke? For another example, do teacher related stress and parent related stress stand out among adolescents who are most likely to smoke? These research questions need to be examined in a relative term with a reference group, and this reference group is naturally the root node in CART. Because the root node represents the average characteristics of the sample (in terms of the independent variables), the characteristics of the subsamples representing the high outcome and low outcome groups can be compared with the average characteristics of the sample. Expressions such as "standing out" come from this comparison. The following section in this chapter continues the discussion on how to determine which independent variables are the most important predictors of the outcome. To discuss well research questions from the within node perspective, it is often necessary to build tables that describe the characteristics of the high outcome and low outcome groups (i.e., terminal nodes). In other words, descriptive statistics such as mean need to be reported for the high outcome and low outcome groups. In the previous chapter, Table 3.5 is designed exactly for this purpose.

From the between node perspective, one should look into at least two issues. CART creates groups (i.e., terminal nodes) with homogeneous cases within each group, but these groups are heterogeneous among themselves. In fact, there is a dynamic in the outcome across the groups. Oftentimes, dramatically different outcomes are present among the groups. This dynamic should be examined. Research questions may pertain to the dramatically varying outcomes among the groups. For example, to what extent does the probability of using tobaccos vary across different segments of adolescents (in the population)? For another example, what is the highest risk or probability of using tobaccos among adolescents? Again, to address research questions like these, some sort of reference is usually preferable. For the second example, the research question is asked often in comparison with the average risk or probability of using tobaccos among adolescents (in the population).

One of the major strengths of CART is its ability to reveal the nonlinear relationships among the data. As alluded to in Chapter 1, complex interaction effects are often very difficult to pinpoint in traditional statistical techniques such as multiple regression analysis. The difficulty increases when the so-called "local interactions" exist (as alluded to in Chapter 2). A typical case is illustrated in Figure 4.1 (see Ma, 2005). This CART analysis is about

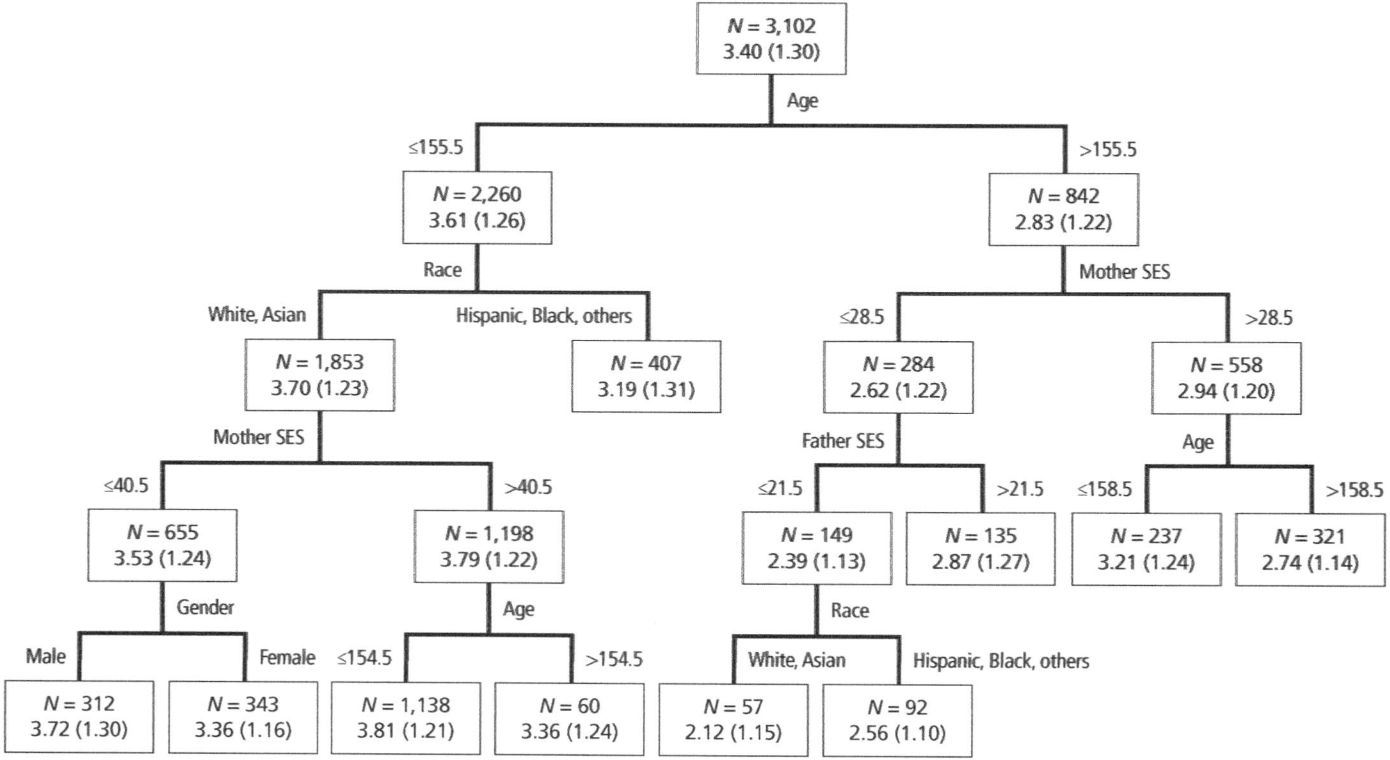

Figure 4.1 CART tree of growth in mathematics achievement during middle and high school, conditional on student background variables. In each node, the top value indicates the number of students, and the bottom value indicates the average rate of growth with standard deviation in parenthesis.

the growth in mathematics achievement during the middle school and high school years based on a national sample of U.S. adolescents. Gender differences in the rate of growth are a very interesting issue in this figure. Traditional multiple regression analysis would indicate a lack of gender differences. Even in the CART analysis, gender differences are essentially absent except hidden in a small corner (see the lower left corner). Gender differences are present for students with lower mother socioeconomic status (SES) but are absent for students with higher mother SES. Based on where the interaction effect shows, this phenomenon is referred to as a *location interaction* because it happens only locally. Because CART reveals meaningfully something that traditional multiple regression cannot reveal, research questions may pertain to the local interactions among the independent variables. A typical research question is whether there are interaction effects that happen only locally within a certain segment of the population. Some more discussion on this issue is provided in an upcoming section in this chapter.

Determining Important Variables

As alluded to in Note 1 of Chapter 1, the labeling of some independent variables as the most important or significant variables is not as straightforward in CART as in, say, multiple regression.[1] The main difficulty is the lack of testing on the relative importance of the independent variables. In traditional multiple regression analysis, one can certainly test one independent variable at a time (examining the *absolute effects* of an independent variable in the absence of other independent variables), but one almost always tests a number of independent variables together (in one model) so that the *relative effects* or *collective effects* of an independent variable can be assessed in the presence of other independent variables. This does not happen in CART because CART tests one independent variable at a time and then selects the one with the largest reduction in a certain impurity measure to partition a parent node. It does not mean that CART ignores the interrelationships among the independent variables. Actually, the CART tree is nothing but relationships among the independent variables (often nonlinear). But the relationships are not established by putting all relevant independent variables together in a relative or collective environment to weigh against one another. CART represents a different philosophy concerning the concept of relationships among the independent variables. Specifically, CART uses the best "performers" in each step of the "job" to establish (often nonlinear) relationships among the independent variables. To accommodate such a philosophy, there may be some different ways to label the most important or significant independent variables.

As mentioned in Chapter 1, from the order in which some independent variables appear in the CART tree (to partition the cases), the term, the best predictors, gets into the literature (e.g., Ture, Kurt, Kurum, & Ozdamar, 2005). It makes a lot of sense to label the first independent variable selected to partition the (whole) data as the best predictor of the outcome (because it is indeed the best candidate to perform the job). The predictability of the independent variables weakens along the order of appearance in the CART tree, and the independent variables that do not appear at all in the CART tree are obviously not an important player in the issue at hand. Overall, under this approach, importance or significance of the independent variables is appreciated in terms of the best (job) performers (i.e., the best predictors) concerning the outcome.

A different approach is proposed in the current book to discuss the importance or significance of the independent variables. As mentioned earlier, the comparison between the characteristics (values) of the independent variables in the high outcome and low outcome groups and the average characteristics (average values) of the same independent variables in the root node reveals reasonably the important independent variables. Any reasonable departure in value of an independent variable in, say, a high outcome group from the average value of the same independent variable in the root node makes the group depart from the population.[2] This independent variable is therefore an important player in separating the group from the population. Importance is defined in this sense. This definition certainly brings back some sense of relativity for the labeling of the important or significant independent variables. Following the statistical convention, one may conduct a test of significance on the difference in the average value of an independent variable between a terminal node and the root node. The test can take the form of one sample t test with the average value of the independent variable in the root node as the (unbiased estimate of the) population parameter. Specifically, for a particular independent variable in a certain terminal node,

$$t = \frac{\bar{x} - m}{s_{\bar{x}}}$$

$$s_{\bar{x}} = \frac{s}{\sqrt{n}}$$

$$df = n - 1$$

where $\bar{x}$ and s are respectively the average value and the standard deviation of the independent variable in the terminal node, n is the sample size of the terminal node (i.e., the node size), and m is the mean of the same independent variable in the root node. Finally, $s_{\bar{x}}$ is the standard error and df is the degree of freedom.

Use Figure 4.1 as an example. The first terminal node in the CART tree occurs at the second level ($N = 407$). A comparison can be made between this terminal node and the root node in terms of the importance of mother SES. For mother SES in this terminal node, $n = 407$, $\bar{x} = 39.73$, and $s = 17.71$. For mother SES in the root node, $m = 41.68$. Using the above formulas, $df = 406$, $s_{\bar{x}} = 20.15$, and $t = -.10$. The result is not statistically significant. Therefore, in terms of mother SES, this terminal node is no different from the root node. In other words, mother SES is not among variables that make this terminal node depart from the population (the root node). In this sense, mother SES is not a statistically significant or important variable.

One may also borrow the concept of effect size (e.g., Cohen's *d*) as a way to discuss the extent of departure associated with the important independent variables. In the above case with all symbols remaining the same in meaning, one can use

$$d = \frac{\bar{x} - m}{s}$$

to calculate Cohen's *d*.[3] The interpretation is straightforward. Cohen's *d* measures the departure in the number of standard deviations, often referred to as the standard deviation units, from the mean of the root node. Cohen's *d* classifies 0.20, 0.50, and 0.80 as small, moderate, and large effect sizes. Follow up with the above example, $d = -.11$ for mother SES in that particular terminal node, indicating obviously trivial importance of mother SES.

Revealing Unique Variables

The way that one examines and reports the behaviors of the independent variables in CART is somewhat different from the way associated with traditional statistical techniques such as multiple regression. One has experienced this uniqueness of CART in the previous section. There are at least two other major perspectives concerning the uniqueness of CART, and the independent variables associated with these perspectives are labeled as unique independent variables and are necessary to get revealed and discussed.

One major perspective is that any independent variables can appear more than once in a CART tree. Going back to Figure 4.1, one can notice that age, mother SES, and race appear more than once. Age actually appears three time in the CART tree. Overall, age is the best predictor of the rate of growth in mathematics achievement during middle and high school years, and perhaps more importantly, age follows the appearance of mother SES to partition the group of students with relatively high mother SES into terminal nodes in two cases. This interesting finding seems to indicate that whenever one talks about the rate of growth in relation to mother SES, one should not fail to mention that there are age differences in the rate of growth among students with higher mother SES. Race also behaves in a unique way. Each time it classifies cases, it classifies White and Asian students into one group and Hispanic, Black, and other students into the other group. The two appearances of race effectively send White and Asian students into both the top and the bottom of the growth spectrum (3.70 and 2.12, respectively, in the rate of growth). Meanwhile, Hispanic, Black, and other students are sandwiched in between (3.19 and 2.56, respectively, in the rate of growth). Therefore, by examining the independent variables that appear more than once in the CART tree, one arrives at a couple of important conclusions. First is the pairing of mother SES and age with age differences in the rate of growth among students with higher mother SES. Second, White and Asian students dominate both the top and the bottom of the growth spectrum (indicating far more variation among students of this group) and Hispanic, Black, and other students dominate the middle of the growth spectrum (indicating far less variation among students of this group). It is worthy of emphasizing that these results cannot be easily, if possibly at all, obtained in traditional statistical techniques such as multiple regression.

The other major perspective concerning the uniqueness of CART pertains to the independent variables that create local interactions. These independent variables often appear only once in a CART tree, but they usually reveal something hidden from the universal point of view. Again, going back to Figure 4.1, one may notice such a case. Gender differences in the rate of growth in mathematics achievement exist only among younger White and Asian students with relatively lower mother SES (among students of high mother SES). Specifically, these White and Asian students are younger than (or as old as) 155.5 months, and their mothers have SES lower than (or equal to) 40.5 which is actually on the high end of mother SES overall.

There may be other unique behaviors of some independent variables in one's CART tree by themselves or in some combinations with other independent variables. Therefore, if one focuses only on the terminal nodes to derive interesting interpretations, much information may be overlooked in the CART tree. To some extent, the careful inspection and examination of a CART tree, by itself, is a "research" effort (process). Stated differently, any CART tree needs to be carefully researched for unique behaviors of some independent variables. Overall, digesting a CART tree is a far more complicated (and often a lot more rewarding) effort than reading off the tables directly on the output from a certain statistical program.

Examining Terminal Nodes

One may easily have a misperception that terminal nodes of a CART tree have measures on the dependent variable that are statistically significantly different from one another. The fact is that although terminal nodes in each branch of the tree do differ statistically significantly from each other, terminal nodes across branches do not necessarily have measures on the dependent variable that are statistically significantly different. Using proper computer algorithms, one can certainly redefine terminal nodes and make these nodes statistically significantly different from one another. But such an idea has never become the goal of a CART analysis. In contrast, it is often considered informative to observe, for example, two terminal nodes with similar measures on the dependent variable sitting on two different branches of the tree. It is informative because, coming from different branches of the tree, the characteristics of cases in those two terminal nodes are dramatically different. This situation often bears important theoretical and practical implications. For example, if one of the terminal nodes represents a group of advantaged students (e.g., socioeconomically advantaged) and the other terminal node represents a group of (socioeconomically) disadvantaged students, there is a good case of resilience when the two terminal nodes share similar learning outcomes of some kind (i.e., similar measures on the dependent variable). Situations like this one are often difficult to expect from conventional statistical techniques such as multiple regression.

Furthermore, the average measure on the dependent variable in a terminal node is not necessarily different statistically significantly from the average measure of the (total) sample on the dependent variable, although it is common for one to compare the node measure with the sample measure when interpreting the results of CART. Using CART, one's primary focus is on the partitioning of parent nodes into child nodes (i.e., one child node average compared with the other child node average). The goal of CART is to make this difference statistically significant. Although the sample average measure can be used as a reference line for interpretation, it never participates in computation in CART. This is not a new idea, however. When gender as an independent variable is included in multiple regression analysis, it is gender differences (i.e., male average compared with female average), rather than male (or female) average compared with the sample average, that are the focus of the investigation.

It is always tempting when examining a terminal node to trace its decision rules (or its partition processes) up to the root node so as to discuss the importance of different independent variables in forming that terminal node (i.e., channeling cases into that terminal node; this issue was discussed in detail in an earlier section). Such an effort is still legitimate and informative from time to time. Nonetheless, from the CART perspective, to gain a better understanding about the dependent variable, a comparison of terminal nodes (i.e., their often dramatically different characteristics on the independent variables as well as their measures on the dependent variable) is more meaningful than a comparison of independent variables for (individual) importance. CART emphasizes characteristics of terminal nodes because it is a statistical procedure to decompose complex interaction effects among independent variables. As a matter of fact, if interactions are built into, for example, a traditional multiple regression analysis, the importance of different independent variables becomes relative as well, in the presence of statistically significant interaction effects. In sum, because of the ability of CART to decompose complex interaction effects, the traditional notion concerning the importance of different independent variables becomes less meaningful (or relevant). When interpreting the results of a CART analysis, one needs to pay close attention to the oftentimes dramatically varying characteristics of cases in different terminal nodes. Documentations on the characteristics of terminal nodes that either share rather similar measures or indicate rather different measures on the dependent variable can become very revealing to the research issues at hand.

Handling Missing Data

Traditional statistical techniques handle missing data by means of deletion (listwise and pairwise) and statistical imputation. These traditional ways of handling missing data are acceptable in a CART analysis, but these procedures often need to be carried out outside of the CART analysis using general statistical packages such as SPSS. CART has its own unique ways of handling missing data. There are discussions on surrogates in the previous chapter. Recall that surrogates of an independent variable chosen to partition a parent node can be used to assign cases with missing data on this independent variable to different child nodes. Therefore, one popular way that CART handles missing data is to use surrogates of the independent variable that has been chosen to partition a parent node.

As the other popular way, CART treats missing data as a valid category. This is unique in that being a new category, missing data can actually take part in partitioning the CART tree. For nominal independent variables, the procedure is straightforward. For example, if an independent variable x originally has two categories (e.g., males and females), with missing data treated as a new category, x has three categories when it enters any CART analysis (males, females, and cases missing on gender).

For continuous independent variables, the procedure is more complex. Consider a simple example in which an independent variable x has three records with valid values (1, 3, 5) and two records with missing values. If one disregards missing data, there are three legitimate binary partitions: (a) 1 versus 3 and 5, (b) 3 versus 1 and 5, and (c) 5 versus 1 and 3. With missing data treated as a new category (and given the notation of M), the following seven binary partitions are all legitimate: (a) 1 versus 3, 5, and M; (b) 1 and M versus 3 and 5, (c) 3 versus 1, 5, and M, (d) 3 and M versus 1 and 5, (e) 5 versus 1, 3, and M, (f) 5 and M versus 1 and 3, and (g) M versus 1, 3, and 5.

This handling of missing data as a new category can be characterized as forcing cases with missing data into the same child node (note that using surrogates to handle missing data is likely to send cases with missing data to both child nodes). Such a treatment of missing data is often called "missing together." This missing together approach is conceptually simple, and one can use this approach to keep track of where cases with missing data are located in a CART tree. This treatment is often sound in many research circumstances. For example, individuals who refuse to provide information on substance use may well be the target group of the research. A good knowledge about cases in the sample is important to take full advantage of this treatment.

Determining Node Size

The issue of an appropriate size for a CART tree has already been discussed in this book. The focus of the previous discussion is mainly data driven (i.e., letting data decide). One sees this point well through the (backward) pruning procedure in which one starts with the minimum node size of 1 case all across (i.e., each case forms a terminal node) and prunes the (huge) tree down progressively. When a desirable tree emerges, pruning stops and the number of levels in the tree is accepted to address the research questions. Some techniques discussed earlier can aid this decision (e.g., validation). Although this is a legitimate approach, the problem one frequently encounters is the existence of rather small terminal nodes. The continuous pruning can "shrink" the number of terminal nodes but may not eliminate trivial terminal nodes. Typically, a terminal node with fewer than five cases can be considered trivial. To avoid or resolve this problem, one usually exercises more control over the tree growth by specifying how many cases a terminal node must have and how many levels a CART tree can grow. This idea prevents the formation of trivial terminal nodes in the first place.

The strategy of specifying sizes for a CART tree and its terminal nodes is in contrast to the above data driven approach. The problem that comes with this strategy is that there is no consensus in the literature regarding the appropriate size of a terminal node. As mentioned earlier, it is safe to define a minimum terminal node as one having fewer than five cases. However, if serious implications for policy and practice are expected from a CART tree, this number (five) surely needs to get bigger. Given that it is a common statistical practice to have at least 50 cases to perform a multiple regression analysis, to define the minimum size of a terminal node as 50 cases is reasonable, in particular when implications for policy and practice are expected from a CART tree. This criterion means that if a partition of a parent node results in one of the child nodes to have fewer than 50 cases, the partition would not be performed so that all terminal nodes in the CART tree would have more than 50 cases each. Of course, such a minimum size of a terminal node implies a large sample (e.g., thousands of cases). Yet, because CART is a data mining technique, it is more powerful to work with large samples than moderate samples (e.g., 200 cases).

The discussion on the size of a terminal node is often related to the size of a CART tree; that is, how many levels should a CART tree have to both capture the essential relationships among independent variables and avoid the overfitting of a CART tree as discussed earlier (i.e., a tree too huge to be meaningful)? Some researchers refer this issue to as the depth of a CART tree. Again, there is no consensus in the literature regarding the appropriate number of levels for a CART tree. It is a common practice to limit the depth of a CART tree to three to five levels for the focus of the analysis and the ease of the interpretation.

Often, one works with both issues (size of a terminal node and depth of a CART tree) together to shape the tree (i.e., control over the tree growth). Most CART software programs allow one to specify the number of levels a CART tree can grow and the number of cases any terminal node must have. With an appropriate sample size, four levels for the tree and 50 cases for each terminal node may be considered reasonable for common CART practices. Of course, a different set of numbers can be proposed and justified for special circumstances (e.g., moderate sample sizes).

Assessing CART Performance

Once one collects data and fits the data to a statistical model, it is often desirable to come up with some indications of how well the data fits the model (often referred to as model data fit). There are some ways to fulfill this purpose in traditional statistical techniques such as multiple regression. One of the most familiar measures is R^2 which indicates the proportion of (total) variance in the dependent variable that has been explained by the model or independent variables in the model. Thus, one often uses R^2 to indicate how well a multiple regression model works. Theoretically in the case of (regular) multiple regression based on ordinal least squares (OLS), R^2 is defined (and calculated) as

$$R^2 = 1 - \frac{SSE}{SST}$$

where *SSE* is the sum of squared errors (residuals) and *SST* is the sum of squares total calculated as the squared differences of the actual values of the dependent variable from their average value (i.e., the mean of the dependent variable).

It is possible to derive a similar measure (R^2) for a CART tree (see the discussion in Chapter 3). In the case of RT (i.e., the dependent variable is continuous), it is straightforward. In fact, the definition of R^2 remains the same with *SSE* as the sum of squared errors (residuals) of the RT tree and *SST* as the sum of squared differences of the values of the dependent variable around its mean in the root node. Some software programs may directly generate what is often referred to as *risk estimate* (see the SPSS Decision Tree program) or *relative error* (see the CART program from Salford Systems). The relative error is the ratio of *SSE* to *SST*. When the risk estimate is available, one can directly use the common notion of variance by squaring the standard deviation in the root node to produce R^2. Back to Figure 4.1 where the CART tree is estimated with SPSS Decision Tree, the risk estimate is the within node variance of the CART tree (equivalent to *SSE*). From the root node, the total variance in the dependent variable can be obtained (equivalent to *SST*). Specifically, the risk estimate = 1.48 and the total variance = 1.69 (i.e., 1.30 × 1.30 in Figure 4.1). $R^2 = (1.69 - 1.48)/1.69 = .12$, indicating that the CART tree accounts for 12% of the variance in the dependent variable. In case that a software program does not provide a measure like risk estimate or relative error, some simple manipulations using, say, SPSS can be carried out for the "manual" calculation. This entails the calculation of the sum of squared differences of the values of the dependent variable around its (root) mean in the root node as *SST*. Then, for each terminal node, one carries out calculation of the sum of squared differences of the values of the dependent variable around its (node) mean in the terminal node. Finally, simply adding this sum across all the terminal nodes produces *SSE*. The interpretation can also be borrowed directly from OLS or multiple regression (i.e., the proportion of total variance in the dependent variable that has been explained by the model or independent variables in the model).

In the case of CT (i.e., the dependent variable is categorical), the goal is to classify cases. One can imagine a "null" (CT) tree that does not use any information from the independent variables to make predictions. As a result, the null tree simply predicts the most popular or common category (in the root node). When a CT tree is established, a question can be asked concerning how much better the CT tree is in making predictions over the null tree. Therefore, a (pseudo) R^2 can be defined (and calculated) as the ratio of the proportion of cases correctly classified by the CT tree to the proportion of the most popular or common category (in the root node). In the case of CT, the risk estimate (from SPSS Decision Tree) indicates the proportion of cases incorrectly classified and thus provides information for the calculation of R^2. In case that a software program does not provide relevant information on this ratio or fraction, manual calculation can be carried out. While the denominator of this fraction is easy to obtain, some manipulations using, say, SPSS are needed for the numerator of this fraction. This entails the categorical coding of each case in a terminal node. When coding information is piled up across all the terminal nodes (as a new variable), this variable can then be compared to the variable with the original categorical information of cases to calculate the proportion of cases correctly classified by the CT tree (case identification is needed to link original and new categories of a case together). The interpretation can use the language of how much better (e.g., 25% better) than the null tree without any independent variables.

Notes

1. Here, the treatment of the root node that provides the key parameters such as the mean, *m*, makes a difference (see Note 2). The formulas so far are based on the conventional statistical procedures assuming that the standard deviation of the population is unknown. This implies that the root node is treated as a sample (from the population). If the root node is considered a population, it makes available not only the mean but also the standard deviation (as the population parameters). In this case, the *t* test can be "downgraded" to the *z* test, and Cohen's *d* can be calculated using the population standard deviation rather than the sample standard deviation.

2. The word *population* is used in an informal way. More precisely, it refers to the population represented by the sample that is the root node. In a boarder sense, however, there are two ways to think of or treat the root node. In a more conservative way, the root node is considered (rightfully) a sample from a well-defined population. The root node, especially when it is a random sample, provides the unbiased estimates of the population parameters. The terminal nodes are internal manipulations within the root node. In a more liberal way, a CART tree may be treated as a self-contained "system" in which the root node can function as a population that generates various terminal nodes. These two treatments do not make any difference if one strictly remains in the CART domain. However, if one intends to bring in some traditional statistical procedures to supplement or extend the CART analysis, the way that the root node is treated may matter (see Note 3).
3. The importance of the independent variables is not necessarily in the order in which the independent variables appear (or are included) in the CART tree. In Figure 4.1, age is used at the root node to partition the rate of growth in mathematics achievement (i.e., age is the most successful independent variable to partition the root node). Another independent variable, father SES, does not appear in the CART tree until it grows to the third level. If a correlation analysis is run among the rate of growth, age, and father SES, then the correlation between the rate of growth and father SES is much stronger than that between the rate of growth and age. Conventionally, father SES is a more important predictor of the rate of growth than age. The traditional notion concerning the importance of different independent variables anchors in the strength of the association between the dependent variable and an independent variable. On the other hand, CART looks for an independent variable that can produce two child nodes that have as much homogeneity in the dependent variable as possible within each child node and as much heterogeneity in the dependent variable as possible between the child nodes.

5

Applications of CART

In this chapter, three applications of CART to some educational issues are discussed as examples of how to formulate research purposes, how to present analytical results, and how to capture important findings in a CART analysis (based on what has been discussed in the previous chapters).[1] Some data to be used are not necessarily current, and as a result, the empirical findings provide historical accounts of relevant educational debates and may not bear direct important implications for educational issues nowadays (even though one application is a published research study). The illustration of CART is the main purpose of using these data. The first application pertains to a RT analysis, and it has been borrowed from time to time in the previous chapter to make a couple of points on CART related issues. The second application pertains to a CT analysis. The final application aims to present a CT analysis incorporating the function of costs and priors with the introduction of profits.

Using Classification and Regression Trees, pages 75–100

Operation of CART Software Programs

There are a few software programs that can be applied to perform a CART analysis, such as C5.0, CART, DTREG, Precision Tree, and SPSS Decision Tree (see Appendix B for more description of each software program). Apart from these "off the shelf" program packages (some refer to them as "point and click" tools), R and Python provide one with the opportunities to write one's own codes to perform a CART analysis. All CART analyses in this book are run with SPSS Decision Tree. In this section, SPSS syntaxes are provided to guide one to perform the three applications to be discussed later on in this chapter.

In SPSS, one can retrieve the Decision Tree command following: Analyze → Classify → Tree. Once the command window opens up, the specification of a CART tree (model) is relatively straightforward. The window operation (point and click) that specifies a CART tree can be translated into a SPSS syntax by using the Paste function (for record keeping). The first application of CART in this chapter categorizes the rate of growth in mathematics achievement during the entire middle and high school years based on student background variables. The dependent variable is the rate of growth in mathematics achievement during the entire middle and high school years and the independent variables are gender, age, race, mother socioeconomic status (SES), father SES, number of parents, and number of siblings. The SPSS syntax for this CART analysis is presented in Appendix C. Although many subcommands take on the default values, one may want to particularly examine a couple of subcommands for the way that some specifications for the CART tree are made.

In the METHOD TYPE subcommand, CRT is the same as CART, and surrogates are used to deal with missing values on the independent variables (i.e., if values of an independent variable are missing then other independent variables highly correlated with the independent variable are used for classification). One can specify a number of surrogates to be made available for use, and the maximum is the number of the independent variables (in a CART tree) minus one (as default). In the GROWTHLIMIT subcommand, the CART tree is (manually) controlled by allowing it to grow four levels with a minimum terminal size of 50 cases (students). In the VALIDATION TYPE subcommand, cross validation is employed to obtain the CART tree. Cross validation divides a sample into a number of folds (i.e., subsamples), and one can specify the number of folds to be created (the maximum is 25). A larger value for the specification on the number of folds indicates a fewer number of cases to be excluded from data analysis during each round of validation. So a CART tree is generated without data from a fold (e.g., the first CART tree is created with all cases except cases from the first fold). Then, the misclassification risk is estimated by applying the CART tree to the fold. The risk estimate (discussed in the previous chapter) for the final CART tree is calculated as the average risk across all CART trees.

One concept or procedure discussed in this book but absent in the syntax (see Appendix C) is pruning. Pruning is not employed in the first application of CART because it would scale back the CART tree severally so that the relationships among the independent variables are not revealed in any meaningful way. When pruning for a CART tree is requested (as it happens in the next chapter), it appears in the METHOD TYPE subcommand as PRUNE=SE(1) (right after the specification of surrogates). Here the (default) value of 1 is the maximum difference in risk expressed in standard errors between the pruned tree and the subtree with the smallest risk (i.e., the 1 SE rule as discussed in Chapter 2). One can increase this value to produce a simpler tree, and one can also set this value to zero to obtain the subtree with the smallest risk.

The second application of CART in this chapter stratifies the sample at hand according to the potential confounding factors to the key variables of interest including cognitive (e.g., achievement) and affective (e.g., attitude) variables. Specifically, this application aims to single out cognitive and affective factors that are associated with whether students take at least precalculus in high school. The potential confounding factors that are taken into consideration include gender, age, race, parental education, parental SES, number of parents, and number of siblings. The SPSS syntax for this CART analysis is almost the same as the one for the first application except for the subcommand of variable specification (i.e., the TREE subcommand). The dependent variable before "BY" and the independent variables after "BY" are different from those in Appendix C. Whether or not students take at least precalculus in high school is the dependent variable and the potential confounding factors listed above are the independent variables. In addition, concerning the subcommand of GROWTHLIMIT, this CART tree is allowed to grow up to five levels (for better stratification of the sample at hand) with the same minimum terminal size of 50 cases (students). Pruning is absent in the second application for better stratification of the sample at hand.

Application 1: Growth in Mathematics Achievement During Middle and High School

In educational research, more and more attention is being paid to the growth rather than the status in learning, as Willets (1988) classically stated that "the very notion of learning implies growth and change" (p. 346). One of the most important educational issues is the growth in academic achievement, in particular in the so-called "core" academic subjects such as mathematics and science. Ma (2005) reported one analysis on growth in mathematics achievement during middle and high school. Data for this analysis come from the Longitudinal Study of American Youth (LSAY), a national, 6-year panel study with a focus on the development of mathematics and science achievement of students in Grades 7 to 12 in the United States (Miller, Kimmel, Hoffer, & Nelson, 2000).

The LSAY employed a stratified random sampling procedure to select 51 public middle and high schools from 12 sampling strata representing geographic region and community type across the United States with probabilities proportional to enrollment. About 60 seventh graders were then randomly selected from each of these schools. These seventh graders were followed for 6 years, from the 1987–1988 school year when they were in Grade 7 to the 1992–1993 school year when they were in Grade 12. The total sample contained 3,116 students. Students wrote mathematics and science achievement tests annually (from Grades 7 to 12), and student, teacher, and principal questionnaires were used to obtain information on characteristics of students and schools.

Using student mathematics achievement measures across the middle and high school grades, Ma (2005) attempted to identify the mechanism (i.e., the interaction among key student background variables) that channels students into groups with differential rates of growth in mathematics achievement during the entire middle and high school years. The analysis proceeds in two stages. In the first stage, hierarchical linear modeling (HLM) techniques are used to set up a growth model that estimates the rate of growth in mathematics achievement for each student (see Raudenbush & Bryk, 2002).

In Ma (2005), the data hierarchy contains repeated measures nested with students (i.e., each student has 6 years of records in mathematics achievement). As a result, the HLM model has two levels. The level one model (within-student model) is a set of separate linear regressions, one for each student. These linear regression equations regress students' scores of mathematics achievement on their grade levels. The intercepts of these linear regression equations are the initial (Grade 7) status of mathematics achievement (because Grade 7 is set as the time zero) and the slopes associated with the time variable, grade level, in these equations are the rate of growth in mathematics achievement. The level one model can be expressed as

$$Y_{it} = \pi_{0i} + \pi_{1i}(grade)_{it} + R_{it}$$

where Y_{it} is the mathematics achievement score for student i at testing occasion t, $(grade)_{it}$ is the grade level that student i is in at testing occasion t, and R_{it} is an error term. As mentioned earlier, the parameters of π_{0i} and π_{1i} represent estimates of the initial status (Grade 7 status) and the rate of growth in mathematics achievement for student i. The level two model contains the between-student regression equation which expresses the rate of growth π_{1i} as

$$\pi_{1i} = \beta_{10} + u_{1i}$$

where the parameter β_{10} is a measure of the average rate of growth in mathematics achievement among students and u_{1i} is an error term (or variance component) that is unique to each student. The individual rates of growth (i.e., π_{1i}) are captured in HLM (see Raudenbush & Bryk, 2002) and are then used as the dependent variable in the second stage of the analysis.

Analysis in the second stage is a CART analysis. The rationale to adopt CART for data analysis is that there is a lack of theoretical insights and empirical studies in regard to growth in mathematics achievement even though growth in mathematic achievement is widely seen as a function of key student background variables and possible interactions among them. Gender (male and female), age, race (Hispanic, Black, White, Asian, and others), mother SES, father SES, number of parents (single parent and both parents), and number of siblings are taken as the key and basic student background variables. The CART analysis, more precisely the RT analysis, is run with these student background variables as the independent variables. The CART analysis is performed with the SPSS Decision Tree software program. To exercise more control over the tree growth, specifications are made to allow the CART tree to grow four levels and to maintain a minimum size of 50 (students) for each terminal node. The results are presented in Figure 5.1.[2] The CART tree in this figure is a part of the SPSS output on a CART analysis. The rest of the SPSS output is presented in Appendix D so that Figure 5.1 and Appendix D together demonstrate a complete SPSS output on a CART analysis.

The root node (i.e., sample) contains 3,102 students (a small number of students are deleted from data analysis due to consistent missing scores on mathematics achievement). The average rate of growth is 3.40 points in mathematics achievement annually. The value in the parenthesis is standard deviation of growth in mathematics achievement. The first partition is done in relation to student age. This indicates that age results in the best

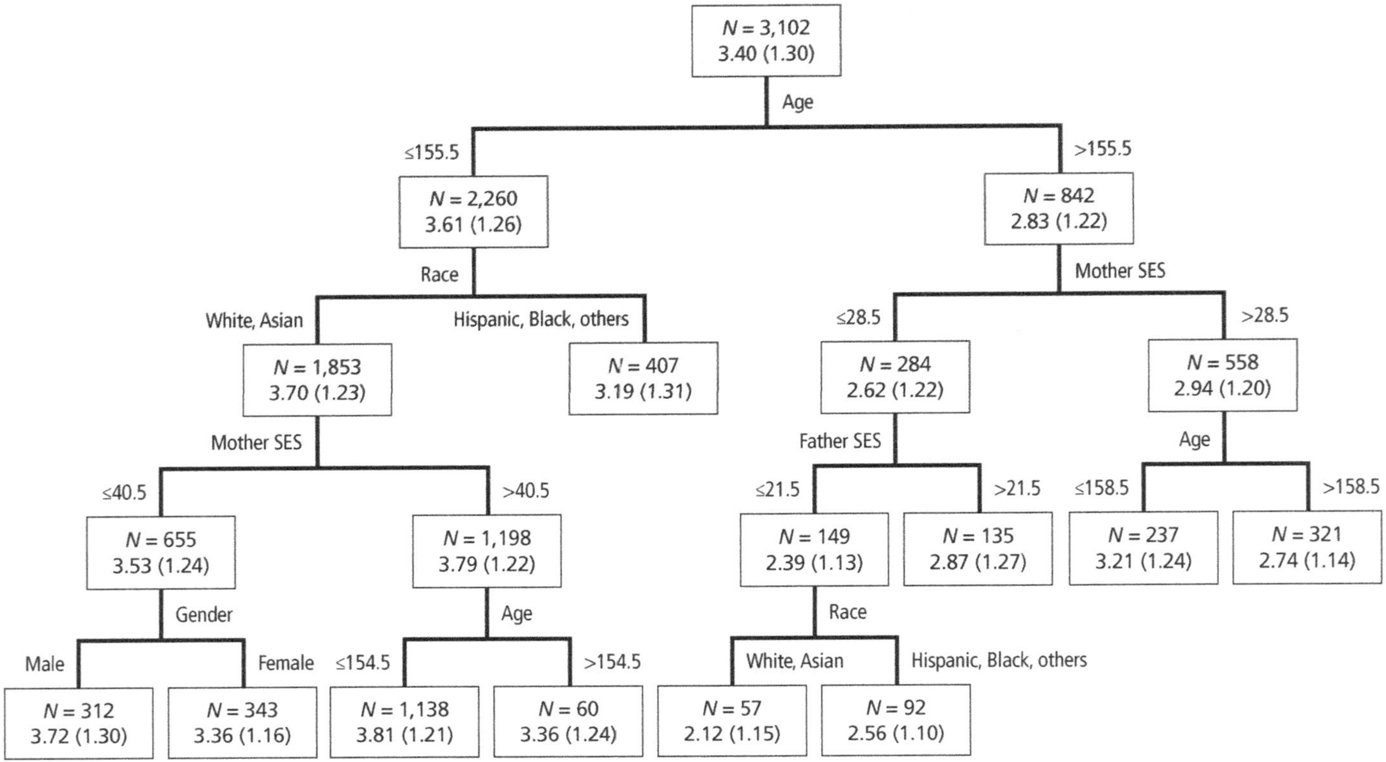

Figure 5.1 CART tree of growth in mathematics achievement during middle and high school, conditional on student background variables. In each node, the top value indicates the number of students, and the bottom value indicates the average rate of growth with standard deviation in parenthesis.

or biggest impurity reduction concerning the rate of growth in mathematics achievement of all independent variables. The left child node contains 2,260 students younger than or as old as 155.5 months, and their rate of growth is 3.61 points each year in mathematics achievement. The right child node contains 842 students older than 155.5 months, and their rate of growth is 2.83 points each year in mathematics achievement.

Among students younger than or as old as 155.5 months, White and Asian students (numbered 1,853) demonstrate a rate of growth at 3.70 points each year in mathematics achievement, while Hispanic, Black, and other students (numbered 407) demonstrate a rate of growth at 3.19 points each year in mathematics achievement. The left child node containing Hispanic, Black, and other students becomes a terminal node. The right child node meanwhile is partitioned according to mother SES. Students with lower mother SES (lower than or equal to 40.5 in the socioeconomic scale) show a rate of growth at 3.53 points each year in mathematics achievement. These 655 students are further partitioned according to their gender. On the other hand, student with higher mother SES (higher than 40.5 in the socioeconomic scale) demonstrate a rate of growth at 3.79 points each year in mathematics achievement. These 1,198 students are further partitioned according to their age.

Among students with lower mother SES, males grow at a rate of 3.72 points each year in mathematics achievement, while females grow at a rate of 3.36 points each year in mathematics achievement. Both child nodes are terminal ones containing 312 and 343 students, respectively. Among students with higher mother SES, those younger than or as old as 154.5 months have a rate of growth at 3.81 points each year in mathematics achievement, while those older than 154.5 months (but younger than or as old as 155.5 months) have a rate of growth at 3.36 points each year in mathematics achievement. Both child nodes are terminal ones with 1,138 and 60 students respectively, and note particularly that the former is the best terminal node with the highest rate of growth in mathematics achievement.

The other side of the CART tree shows partitions of students older than 155.5 months. The 284 students with lower mother SES (lower than or equal to 28.5 in the socioeconomic scale) grow at a rate of 2.62 points each year in mathematics achievement, while the 558 students with higher mother SES (higher than 28.5 in the socioeconomic scale) grow at a rate of 2.94 points each year in mathematics achievement. Both child nodes become parent ones for further partitions.

Among students with lower mother SES, those 135 students with higher father SES (higher than 21.5 in the socioeconomic scale) form a terminal node with a rate of growth at 2.87 points each year in mathematics achievement, while those 149 students with lower father SES (lower than or equal to 21.5 in the socioeconomic scale) are further partitioned into two terminal nodes according to their race. The 57 White and Asian students form a terminal node with a rate of growth at 2.12 points each year in mathematics achievement. The 92 Hispanic, Black, and other students form another terminal node with a rate of growth at 2.56 points each year in mathematics achievement. Note that the former is the worst terminal node with the lowest rate of growth in mathematics achievement.

Going back two levels of the CART tree, one sees that among the 558 students with higher mother SES (higher than 28.5 in the socioeconomic scale), those younger than or as old as 158.5 months (but older than 155.5 months) form a terminal node with a rate of growth at 3.21 points each year in mathematics achievement, while those older than 158.5 months form another terminal node with a rate of growth at 2.74 points each year in mathematics achievement. These two terminal nodes contain 237 and 321 students, respectively.

The CART analysis has revealed a wide range of growth in mathematics achievement with the rate of growth ranging from 2.12 to 3.81 points each year in mathematics achievement. Table 5.1 describes the background characteristics of students in each of the terminal nodes that are arranged in rate of growth from low (G1) to high (G10) with the first node (G0) as the root node. Descriptive statistics show that for the terminal node with the highest rate of growth in mathematics achievement (G10), gender is almost balanced with 53% of students being female in that node. Students are predominantly White with 96% being White and 4% being Asian (Hispanic, Black, and other students are absent). Students in this node average the highest father SES and the second highest mother SES across all terminal nodes. These students are the youngest among all terminal nodes, and they have the fewest number of siblings. Therefore, this terminal node with the best rate of growth in mathematics achievement portrays an equal number of males and females who are predominantly White, are the youngest in the student population (the same grade cohort), and come from wealthy families with adequate attention from parents (due to fewer siblings).

On the other hand, students in the terminal node with the worst rate of growth in mathematics achievement are mostly males with 30% being females and are predominantly White with 95% of the students being White and 5% being Asian (Hispanic, Black, and other students are absent). Students in this node average both the lowest father SES and the lowest mother SES across all terminal nodes. These students form the second oldest group (node) among all terminal nodes, and they have the largest number

TABLE 5.1 Means of Rates of Growth and Student Background Characteristics in Terminal Groups

	G0 (3,102)	G1 (57)	G2 (92)	G3 (321)	G4 (135)	G5 (407)	G6 (237)	G7 (60)	G8 (343)	G9 (312)	G10 (1,138)
Growth (–1.25–8.84)	3.40	2.12	2.56	2.74	2.87	3.19	3.21	3.36	3.36	3.72	3.81
Female (in proportion)	0.48	0.30	0.25	0.29	0.31	0.53	0.40	0.44	1.00	0.00	0.53
Hispanic (in proportion)	0.10	0.00	0.75	0.08	0.10	0.44	0.05	0.00	0.00	0.00	0.00
Black (in proportion)	0.12	0.00	0.25	0.20	0.18	0.50	0.07	0.00	0.00	0.00	0.00
White (in proportion)	0.73	0.95	0.00	0.67	0.67	0.00	0.85	0.95	0.98	0.97	0.96
Asian (in proportion)	0.04	0.05	0.00	0.02	0.03	0.00	0.01	0.05	0.02	0.03	0.04
Others (in proportion)	0.01	0.00	0.00	0.03	0.02	0.06	0.02	0.00	0.00	0.00	0.00
Mother SES (12.00–88.00)	41.68	21.22	19.75	47.52	20.38	39.73	49.75	54.58	26.79	27.39	53.92
Father SES (12.00–89.00)	41.53	18.18	18.60	37.59	41.28	37.42	43.37	46.71	37.13	37.26	48.48
Age (103.00–195.00)	152.83	163.08	163.70	164.30	161.96	148.93	156.79	155.00	149.50	149.11	148.66
Siblings (1.00–9.00)	2.88	3.29	3.11	2.88	3.03	2.93	2.74	2.74	2.82	2.89	2.64

Note: Numbers in parentheses under group identifications are sample sizes. Numerical values in other parentheses indicate ranges (i.e., minimum and maximum). SES = socioeconomic status. Unit for age is month.

of siblings. These findings present a picture of predominantly White males who are among the oldest in the student population (the same grade cohort), come from low-income families, and very likely have inadequate attention from parents (with the largest number of siblings at home).

The CART analysis also has revealed quite a few interesting findings that are relevant to educational policies and practices. For example, what is the role of race in student growth in mathematics achievement? In Figure 5.1, one sees that race interacts directly with age among younger students (younger than or as old as 155.5 months), and race also interacts directly with father SES among older students (older than 155.5 months). In both cases, White and Asian students demonstrate similar behaviors on growth, while Hispanic, Black, and other students demonstrate similar behaviors on growth.

Among younger students, White and Asian students represent a (parent) node that has one of the best rates of growth in mathematics achievement (3.70 points each year), and as a matter of fact, this group of White and Asian students descends the terminal node that has the highest rate of growth in mathematics achievement. In contrast, among older students, White and Asian students form the terminal node that has the lowest rate of growth in mathematics achievement (2.12 points each year). Therefore, younger White and Asian students grow at the best rate in mathematics achievement, but older White and Asian students with low mother and father SES grow at the worst rate in mathematics achievement. This polarization of White and Asian students in the rate of growth in mathematics achievement, though very much imbalanced with substantially more fast-growing White and Asian students than slow-growing White and Asian counterparts, has rarely been documented in previous research studies.

Hispanic, Black, and other students are sandwiched in between. The two terminal nodes with Hispanic, Black, and other students do not have rates of growth as dramatically different as those for White and Asian students. Younger Hispanic, Black, and other students grow at a rate of 3.19 points each year in mathematics achievement, while older Hispanic, Black, and other students with low mother and father SES grow at a rate of 2.56 points each year in mathematics achievement. The phenomenon of polarization in the rate of growth in mathematics achievement as evidenced for White and Asian students is absent for Hispanic, Black, and other students. These findings indicate that age and parental SES have substantially more impacts on the rate of growth in mathematics achievement for White and Asian students than for Hispanic, Black, and other students.

Another interesting finding that has been alluded to in the previous chapter pertains to the local gender differences in the rate of growth in mathematics achievement. In the current CART analysis, gender differences occur locally in a small corner (see the lower left corner in Figure 5.1). In fact, gender differences are present only for students with lower mother SES but are absent for students with higher mother SES. Again, this phenomenon is referred to as a local interaction (between gender and mother SES) because it happens only locally.

Finally, the previous chapter has already used Figure 5.1 (same as Figure 4.1) to discuss the issues of important independent variables and the proportion of variance in the dependent variable explained by the independent variables (i.e., R^2). As a brief review, one can examine another independent variable for significance. In Table 5.1, Group 1 (G1) has the slowest rate of growth in mathematics achievement during the entire middle and high school years. This group has father SES as 18.18 (i.e., $\bar{x}$ = 18.18), $n = 57$ (so that $\sqrt{n}$ = 7.55), and $s = 2.13$ (so that $s_{\bar{x}}$ = 2.13/7.55 = .28). For father SES in the root node, $m = 41.53$. Therefore, t = (18.18 – 41.53)/.28 = –83.39 (df = 56). The result is statistically significant, indicating that, in terms of father SES, this terminal node is significantly different from the root node. With d = (18.18 – 41.53)/2.13 = –10.96, effect size indicates a large effect (in terms of absolute value). Therefore, father SES is a statistically significant or important variable for this terminal node. In other words, father SES is an independent variable that makes this terminal node depart significantly from the population (the root node). In Appendix D, the risk estimate = 1.49 for the CART tree. In Figure 5.1, the total variance = 1.30 × 1.30 = 1.69 in the root node. R^2 = (1.69 – 1.48)/1.69 = .12, indicating that the CART tree explains 12% of the variance in the dependent variable (i.e., the rate of growth in mathematics achievement during the entire middle and high school years).

Application 2: Dropping Out of Advanced Mathematics in Middle and High School

There have been many concerns about mathematics education in the United States. One of them is the disproportionate number of high school students who drop out of the study of mathematics, particularly advanced mathematics, prematurely. As National Council of Teachers of Mathematics (2000) warned a long time ago, this problem bears significant individual and social consequences as the global economy demands a more and more mathematically literate workforce. Using the LSAY data, this application concerns the issue of participation in the most advanced mathematics courses in high school (i.e., precalculus and calculus).

The LSAY data contain detailed information on mathematics courses that students take in each year of their middle and high school (Grades 7 to 12). This application aims to single out cognitive (e.g., achievement) and affective (e.g., attitude) factors that are associated with whether students take at least precalculus in high school (i.e., the probability that students take at least precalculus in high school). Specifically, with longitudinal data over the entire middle and high school (Grades 7 to 12), one can study how the extent to which students progress (either positively or negatively) in both cognitive and affective domains in mathematics education (i.e., the rates of change in both cognitive and affective factors) is associated with the probability that students take at least precalculus in high school. In the cognitive domain, included in the analysis are the rate of change in overall achievement in mathematics and the rates of change in achievement in basic skills, algebra, geometry, and quantitative literacy. In the affective domain, included in the analysis are the rates of change in attitude toward mathematics, mathematics anxiety, and self-esteem.

An attempt is made to take into account potential confounding factors that can distort the relationship between the probability that students take at least precalculus and the rates of change in cognitive and affective impact factors. Gender, age, race, parental (mother and father) education level, parental SES, family structure (either both-parent families or single-parent families), and number of siblings are considered as the potential confounding factors.

Methodologically, this application adopted the analytic model outlined in Zhang and Bracken (1996) who presented a risk-factor analysis using tree-based stratification. This analysis includes two steps. The first step is to stratify the sample according to the potential confounding factors, and the second step is to calculate the effects of the putative risk factors adjusted for confounders through sample stratification. Zhang and Bracken (1996) emphasized that the overall goal of this type of analysis is to use tree-based methods, such as CART, to reduce the data dimension of the confounders and to build a filter for the assessment of the putative risk factors.

Table 5.2 presents the descriptive information on both the putative risk factors and the potential confounding factors. In the first step of the analysis, potential confounding factors are used to stratify the sample on the basis of CART. Specifically, the CART analysis partitions the sample into a number of homogeneous terminal nodes (see Figure 5.2). The root node contains 3,116 students (the original sample size of the LSAY). The overall probability of taking at least precalculus in high school is 21.34%. Mother education level produces the best or biggest impurity reduction among all potential confounding factors in this root node, dividing it into two child nodes. One

TABLE 5.2 Descriptive Statistics of Putative Impact Factors and Potential Confounding Factors

	Minimum	Maximum	Mean	SD
Putative Cognitive Impact Factors				
Rate of Change in Overall Achievement in Mathematics (Continuous)	–1.15	8.74	3.48	1.31
Rate of Change in Achievement in Basic Skills (Continuous)	–0.59	7.92	3.69	1.22
Rate of Change in Achievement in Algebra (Continuous)	1.92	16.26	7.66	3.07
Rate of Change in Achievement in Geometry (Continuous)	0.22	14.04	6.21	2.65
Rate of Change in Achievement in Quantitative Literacy (Continuous)	–0.21	7.52	3.83	1.37
Putative Affective Impact Factors				
Rate of Change in Attitude Toward Mathematics (Continuous)	–1.29	0.60	–0.28	0.22
Rate of Change in Mathematics Anxiety (Continuous)	–0.25	0.39	0.05	0.08
Rate of Change in Self-Esteem (Continuous)	–1.40	1.48	0.10	0.31
Potential Confounding Factors				
Gender (Categorical, 1 = Female and 2 = Male)				
Age (Continuous)	103.00	195.00	152.83	7.83
Race (Categorical, 1 = Hispanic, 2 = Black, 3 = White, 4 = Asian, 5 = Others)				
Mother Education Level (Continuous)	10.00	18.00	12.70	2.09
Father Education Level (Continuous)	10.00	18.00	13.27	2.50
Mother Socioeconomic Status (Continuous)	12.00	88.00	41.68	16.96
Father Socioeconomic Status (Continuous)	12.00	89.00	41.53	20.62
Family Structure (Categorical, 1 = Both-Parent and 2 = Single-Parent)				
Number of Siblings (Continuous)	1.00	9.00	2.51	1.15

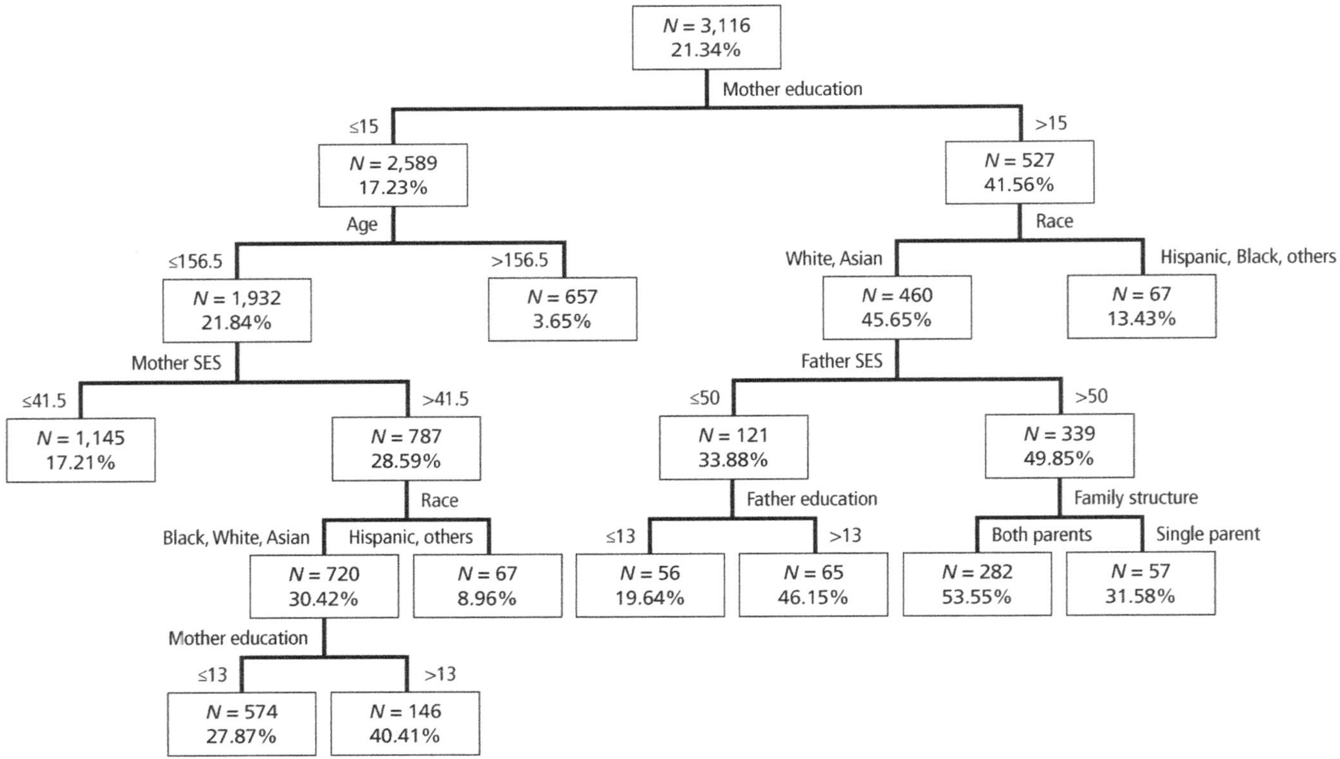

Figure 5.2 CART tree of participation in the most advanced mathematics coursework (pre-calculus or calculus) in high school, conditional on student background variables. In each node, the top value indicates the number of students, and the bottom value indicates the probability or proportion of students taking at least pre-calculus in high school.

node contains 2,589 students with mother education level less than or equal to 15 years, and the other contains 527 students with mother education level more than 15 years. The probability of taking at least precalculus in high school is 17.23% and 41.56%, respectively, for the two child nodes.

The left node then becomes the parent one of two age child nodes (younger than or as old as 156.5 months; older than 156.5 months). The 657 older students form a terminal node with a probability of taking at least precalculus in high school being 3.65%. The 1,932 younger students with a probability of taking at least precalculus in high school being 21.84% form a parent node that is divided into two child nodes based on mother SES (lower than or equal to 41.5 and higher than 41.5 on the socioeconomic scale). The left child node becomes a terminal node of 1,145 students with a probability of taking at least precalculus in high school being 17.21%. The right child node with a probability of taking at least precalculus in high school being 28.59% becomes a parent node that descends two racial child nodes (Black, White, and Asian; Hispanic and others). The 67 students with Hispanic and other racial backgrounds form a terminal node with a probability of taking at least precalculus in high school being 8.96%. The 720 Black, White, and Asian students with a probability of taking at least precalculus in high school being 30.42% are further partitioned into two terminal nodes according to, again, mother education level. The 574 students with mother education level less than or equal to 13 years have a probability of taking at least precalculus in high school at 27.87%, whereas the 146 students with mother education level more than 13 years (but less than or equal to 15 years) have a probability of taking at least precalculus in high school at 40.41%.

The other side of the CART tree structure shows a partition of students with mother education level more than 15 years into two racial child nodes. The 67 students with Black, Hispanic, and other racial backgrounds form a terminal node with a probability of taking at least precalculus in high school being 13.43%. The 460 White and Asian students with a probability of taking at least precalculus in high school being 45.65% are divided into two child nodes according to father SES (lower than or equal to 50 and higher than 50 on the socioeconomic scale). The left child node (121 students) with a probability of taking at least precalculus in high school being 33.88% further descends two terminal nodes based on father education level. The 56 students with father education less than or equal to 13 years have a probability of taking at least precalculus in high school at 19.64%, whereas the 65 students with father education more than 13 years have a probability of taking at least precalculus in high school at 46.15%. The right child node (339 students) with a probability of taking at least precalculus in high school being 49.85% is divided into two terminal nodes according to family structure. The 282 students from both-parent families show a probability of taking at least precalculus in high school at 53.55%, whereas the 57 students from single-parent families show a probability of taking at least precalculus in high school at 31.58%.

As one can see in Figure 5.2, the ten terminal nodes demonstrate dramatically different probabilities of taking at least precalculus in high school. These probabilities range from 3.65% to 53.55%. Not only is this CART analysis quite revealing in itself, but also it serves to identify ten terminal nodes for sample stratification. Because each student falls into one of these terminal nodes, these 10 terminal nodes define 10 strata for the entire sample.

In the second step, a series of logistic regression analyses are carried out. Following Zhang and Bracken (1996), potential confounding factors that appear in the CART tree (see Figure 5.2) are included in a logistic regression analysis, including age, race (as a dichotomous variable), mother education level, father education level, mother SES, father SES, and family structure. These confounding factors have shown main and second-order interaction effects in the CART tree that stratifies the sample, and they are entered into the logistic regression in a forward stepwise manner. Father SES is removed from the equation because of insignificance, and the remaining factors form a base model. Each putative impact factor is then entered into this base model so that the effect of this putative impact factor on the probability of taking at least precalculus in high school can be adjusted by those stratifying variables (confounding factors) in this base model.

Table 5.3 presents the adjusted effects of the rates of change in cognitive and affective factors on the probability of taking at least precalculus in high school.[3] All five rates of change in cognitive factors have statistically significant effects. Students who grow fast in (overall) mathematics achievement are more than 2.5 times as likely to take at least precalculus in high school as students who grow slow in mathematics achievement. Examining different areas of mathematics, one sees that faster rates of growth in basic skills and quantitative literacy increase the probability (nearly 2.5 times) of taking at least precalculus in high school. In comparison, the rates of growth in algebra and geometry are less important to taking the most advanced mathematics courses in high school.

Interesting findings appear regarding the effects of the rates of change in affective domains. The rate of change is not related to the probability of taking at least precalculus in high school for either mathematics anxiety or self-esteem. However, among all putative cognitive and affective factors, the rate of change in attitude toward mathematics turns out to be the most important

TABLE 5.3 Adjusted Effects of Changes in Cognitive and Affective Factors During Middle and High School on Participation in Advanced Mathematics Courses (Pre-Calculus or Calculus) in High School

Factor of Change	Effect	SE	Exp	95% CI
Cognitive Factors				
Rate of Change in Overall Achievement in Mathematics	0.99*	0.07	2.70	2.35–3.10
Rate of Change in Achievement in Basic Skills	0.83*	0.07	2.29	2.00–2.62
Rate of Change in Achievement in Algebra	0.54*	0.03	1.72	1.61–1.83
Rate of Change in Achievement in Geometry	0.57*	0.03	1.77	1.65–1.89
Rate of Change in Achievement in Quantitative Literacy	0.90*	0.06	2.45	2.17–2.77
Affective Factors				
Rate of Change in Attitude Toward Mathematics	1.49*	0.28	4.44	2.57–7.68
Rate of Change in Mathematics Anxiety	0.51	0.78	1.67	0.36–7.69
Rate of Change in Self-Esteem	0.33	0.19	1.39	0.96–2.02

* $p < .05$

Note: SE denotes standard errors. Exp denotes the regression results in terms of *e* raised to the power of each effect. CI denotes confidence interval of each Exp.

factor that influences the probability of taking at least precalculus in high school, far more important than the rates of change in cognitive factors. Specifically, students who have favorable rates of change in attitude are almost 4.5 times more likely to take at least precalculus in high school than students who have unfavorable rates of change in attitude. Linking with the rates of change in cognitive factors, one can conclude that although the rates of change in cognitive factors have more comprehensive effects than the rates of change in affective factors on the probability of taking at least precalculus in high school, it is the rate of change in attitude, an affective factor, that is most predictive of the probability of taking at least precalculus in high school.

Using CART to create data strata for data analysis of a critical issue at hand is a good research strategy in that it makes rather pure or refined implications for policy and practice. Because of the stratification that separates students into homogeneous categories of behaviors, the effects of the key independent variables of interest are adjusted over those independent variables that create the strata. Implications for policy and practice can then be made with a much more confident and valid adjustment over potential confounding factors. This adjustment no doubt increases the chance of success for any treatments or interventions that are developed based on the specific implications.

Application 3: Science Coursework Among Tenth Graders in High School

The current application continues the substantive theme from the previous application. Contemporary education witnesses from time to time concerns about school coursework in what is often referred to as the core content areas, one of which is science. For example, the American Association of State Colleges and Universities (2006) discussed a wave of national attention aimed at beefing up the high school curriculum and why the postsecondary education community must join this national effort as a partner. This application of CART offers an exploratory opportunity to examine how high school students (i.e., the tenth graders) take important science courses in the United Kingdom. The sample of tenth graders ($N = 14{,}157$) in the United Kingdom comes from the 2015 Programme for International Student Assessment (PISA). The 2015 assessment cycle focuses on science education.

Information is available on whether students take two important science courses, physics and chemistry, in Grade 10 in the United Kingdom. A variable of science coursework is created as the dependent measure with 1 = none (i.e., neither physics nor chemistry), 2 = physics (only), 3 = chemistry (only), and 4 = both (i.e., both physics and chemistry). The independent variables include students' age (continuous), gender (male, female), immigration status (native, immigrant), father SES (continuous), and mother SES (continuous). The goal of this exploratory analysis is to examine how students with these individual background characteristics enroll in physics and chemistry in Grade 10. A CART analysis (more precisely CT because the dependent variable is categorical) can be carried out.

The discussion in Chapter 3 on using costs and priors for misclassification bears some useful meanings in this application. Without any specification of costs or priors, any misclassification has the same consequence on tree growth (i.e., misclassification costs the same). There are circumstances in research where misclassification may not cost the same. In this application, misclassifying a student without either course into the category of students with both courses is obviously a bigger mistake than misclassifying the same student into the category of students with either physics or chemistry. Similarly, misclassifying a student with both courses into the category of students without either course is obviously a bigger mistake than misclassifying the same student into the category of students with either physics or chemistry. As discussed in Chapter 3, such concerns can be incorporated into a CT analysis. Table 5.4 presents the specification of costs for misclassification. One can see that misclassifying a student without either course into the category of students with both courses costs twice as much (the

TABLE 5.4 Specification of Costs

	Predicted			
Actual	**None**	**Physics**	**Chemistry**	**Both**
None		1	1	2
Physics	1		1	1
Chemistry	1	1		1
Both	2	1	1	

value of 2) as misclassifying the same student into the category of students with either physics or chemistry (the value of 1). Meanwhile, misclassifying a student with both courses into the category of students without either course costs twice as much as misclassifying the same student into the category of students with either physics or chemistry. Finally, misclassifying a student with physics (chemistry) into the category of students with chemistry (physics) has a cost value of 1, and misclassifying a student with either physics or chemistry into the category of students with both courses also has a cost value of 1. Appendix E contains the SPSS Decision Tree syntax that incorporates this specification of costs (see the COSTS CUSTOM subcommand). Under this subcommand, with the coding of courses presented earlier, "1 1 [0]" means that "none" into "none" is correct (i.e., it costs 0), "1 2 [1]" and "1 3 [1]" indicate that "none" into either "physics" or "chemistry" costs 1, and "1 4 [2]" means that "none" into "both" costs 2, just to name the first few.

The syntax in Appendix E also contains another subcommand PROFITS CUSTOM with the pattern of "1 [0 0] 2 [2 1] 3 [2 1] 4 [4 2]" which represents a simple but useful idea of profits. This subcommand corresponds to Table 5.5 in which there is a specification of profits for each category of the dependent variable. Specifically, taking none of the courses generates no revenue (return) and demands no expense (effort) (i.e., "1 [0 0]"), and profits are calculated automatically (by SPSS) as revenue minus expense (0 in this case). Taking either physics or chemistry demands effort of 1 and generates return of 2 (i.e., "2 [2 1]" or "3 [2 1]"). So, profits are 1 for

TABLE 5.5 Specification of Profits

	Revenue	**Expense**	**Profit**
None	0	0	0
Physics	2	1	1
Chemistry	2	1	1
Both	4	2	2

taking either physics or chemistry. Finally, taking both courses demands double effort but also generates double return (i.e., "4 [4 2]"). So, profits are 2 for taking both courses.

The function of profits does not influence the tree growth but adds an interesting aspect to each terminal node. Some terminal nodes profit more than other terminal nodes. To some extent, profits function like adding another (outcome) measure apart from the dependent variable to describe a terminal node. As a result, there is more information about a terminal node in terms of its outcome, and comparisons among terminal nodes also become more informative. When some sort of income is present, the terms of revenue and expense have monetary meanings. In research cases where there is not any real income, revenue and expense can be given specific meanings based on the research issue. In the case of taking science courses in this application, it is logical to think of revenue as return (of education) that helps students excel in science education (e.g., performing better in science tests) and expense as effort that students need to invest to complete the coursework. Overall, to apply the function of profits, innovative meanings can be sought and attached to categories of a dependent variable.

One can run a CT analysis with the above specifications on both costs and profits. Figure 5.3 depicts the left branch of the CT tree. Because data analysis is carried out on a nationally representative sample, the results from the root node and each terminal node are generalizable to the population. The root node shows that 63.78% of the population of the tenth graders (in the United Kingdom) take both physics and chemistry while 20.66% of the population take neither physics nor chemistry. Meanwhile, 7.06% take physics only, and 8.51% take chemistry only. The left branch of the CT tree heavily emphasizes father SES (appearing twice) and especially mother SES (appearing four times). A local gender gap (i.e., gender differences) is also detected. The right branch of the CT tree (see Figure 5.4) is entirely "dominated" by mother SES (appearing four times). Overall, the CT analysis highlights mother SES as the single critical "force" behind the way that the tenth graders take physics and chemistry (among age, gender, immigration status, father SES, and mother SES). Indeed, mother SES is the single most important variable responsible for the way that the tenth graders take physics and chemistry.

The whole CT tree (Figures 5.3 and 5.4) contains 11 terminal nodes (groups). The pattern of their differences can be appreciated from different perspectives. In general, a rather positive picture emerges concerning

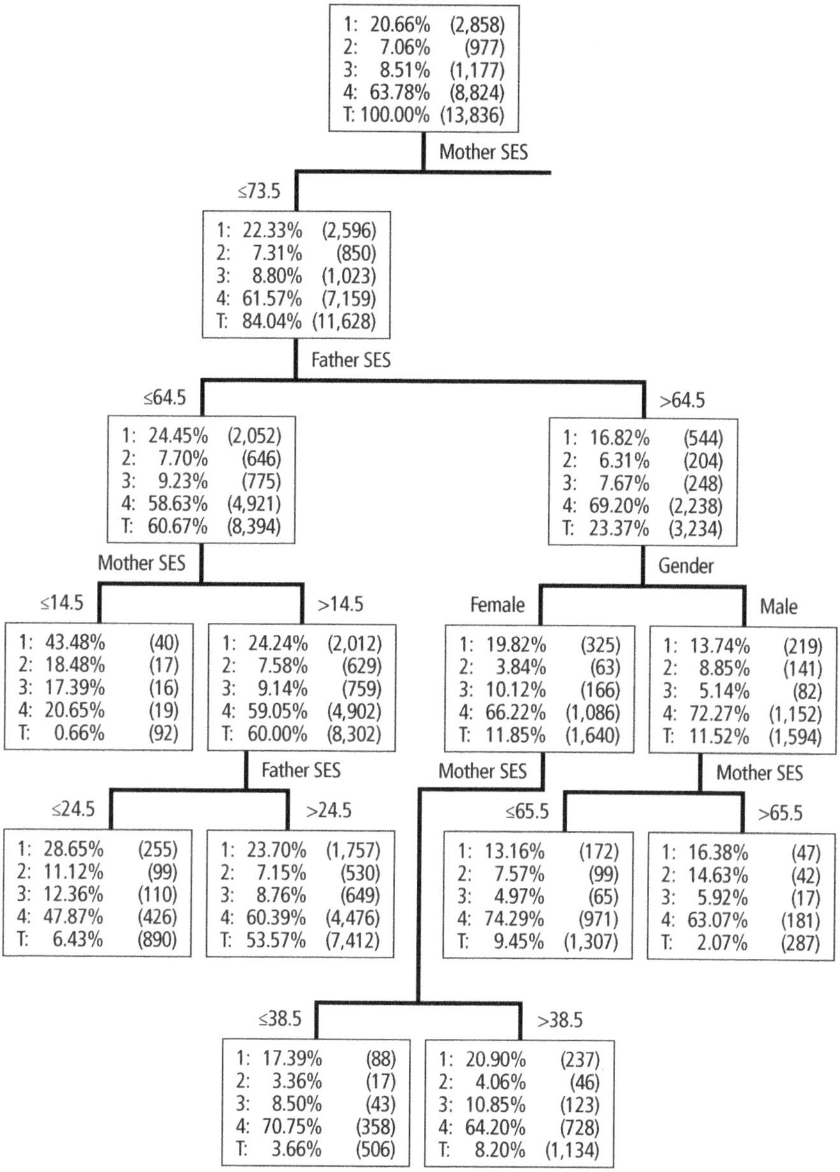

Figure 5.3 Partial (left) CT tree on tenth grade students taking science courses. In each box, the first column indicates categories with 1 = None, 2 = Physics, 3 = Chemistry, 4 = Both, and T = total. The percentage and the number of students of each category follow.

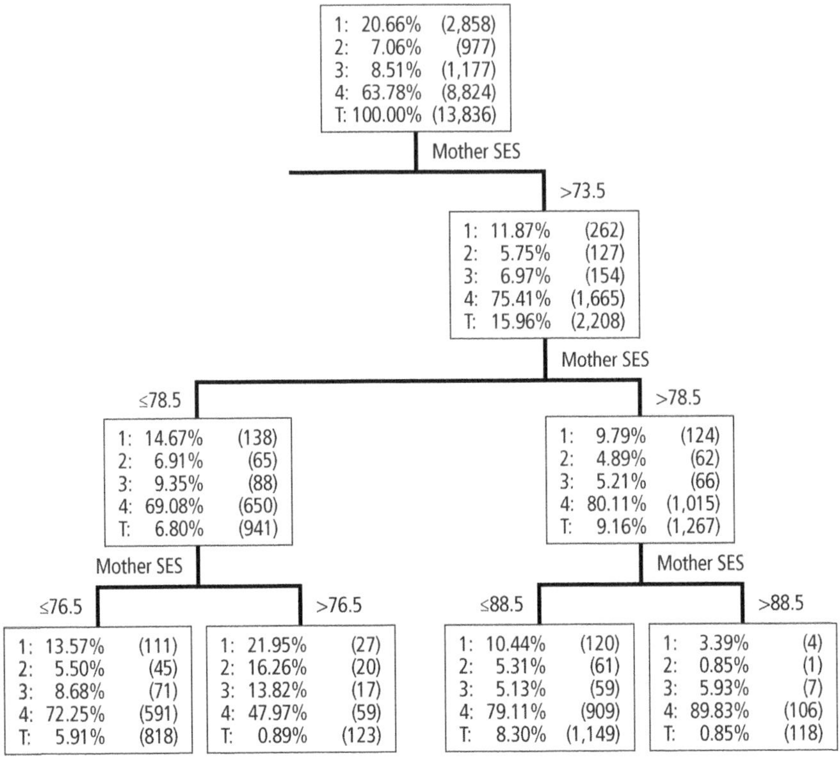

Figure 5.4 Partial (right) CT tree on tenth grade students taking science courses. In each box, the first column indicates categories with 1 = None, 2 = Physics, 3 = Chemistry, 4 = Both, and T = total. The percentage and the number of students of each category follow.

the behaviors of the tenth graders taking physics and chemistry. The percentage of the tenth graders taking both physics and chemistry ranges from 20.65 to 89.83, but only three out of the 11 groups have a percentage below 50. These three groups represent only 0.66 + 6.43 + 0.89 = 7.98% of the population. The percentage of the tenth graders taking neither physics nor chemistry ranges from 3.39 to 43.48. There are only two groups where more than one in four tenth graders take neither physics nor chemistry, representing only 0.66 + 6.43 = 7.09% of the population. When it comes to taking either physics or chemistry, there are more tenth graders preferring chemistry over physics in six out of the 11 groups. In each of the 11 groups, the sum of the percentages of the tenth graders taking physics alone and chemistry alone is substantially less than the percentage of the tenth graders taking both physics and chemistry. There is only one exception for the group with 92 tenth graders. Given such a tiny group (representing only 0.66% of the population), the aforementioned pattern is overwhelming. Finally, a very unique phenomenon can be noticed in Figures 5.3 and 5.4. Apart from one group that represents 53.57% of the population of the tenth graders, all other (10) groups represent less than 10% of the population. This phenomenon signals a wide range of "local" behaviors of the tenth graders taking physics and chemistry that are so different from the "mainstream" behaviors of the tenth graders taking physics and chemistry. To some extent, one's understanding of the mainstream behaviors of the tenth graders taking physics and chemistry can be quite misleading because nearly half of the population demonstrates different local behaviors of taking physics and chemistry.

To compare the characteristics of the tenth graders across terminal nodes (groups), a table of descriptive statistics on the independent variables in a group by group format is one of the best ways (as discussed in the previous chapters). Table 5.6 is such a table. One can see clearly that both age and immigration status are very similar across the 11 groups. Meanwhile, single gender groups are formed only locally (concerning four groups). Yet, father SES and mother SES vary substantially across the 11 groups. Although father SES varies in a wide range from 17.80 to 75.34, mother SES varies even more from 13.67 to 89.00. Specifically as an example, the tenth graders in the first group ($N = 92$) are 15.69 years in age, 41% of them are male and 12% of them are immigrants. Their father SES is 30.27 (the second lowest among the groups) and their mother SES is 13.67 (the lowest among the groups). This group is at the highest risk of inadequately preparing its members in the tenth grade science coursework. The discussion on each group of the tenth graders becomes much more meaningful when compared with the sample (i.e., the population in this case). For this reason, Table 5.6 contains descriptive information of the population to facilitate any comparison. The aforementioned group is not much different from the population in terms of age (15.69 versus 15.72) and immigration status (0.12 versus 0.13). The group has some less male presence compared with the population (0.41 versus 0.51). However, it has much lower father SES (30.27 versus 50.08) and in particular mother SES (13.67 versus 49.91) than the population.

Some significance tests may be carried out if one is especially interested in the special effects of some independent variables in certain terminal nodes (groups). As discussed in Chapter 4, the major motivation is to find out what independent variables make one group depart significantly from the population. With the first group as the example again, a t test of mother SES is statistically significant (see Chapter 4 for formulas). Specifically, standard error is

TABLE 5.6 Group Characteristics of Tenth Graders Taking Physics and Chemistry

Group	*N*	Age		Male		Immigrant		Father SES		Mother SES	
		Mean	SD	Mean	SD	Mean	SD	Mean	SD	Mean	SD
1	92	15.69	0.28	0.41	0.50	0.12	0.33	30.27	12.08	13.67	0.80
2	818	15.72	0.29	0.51	0.50	0.12	0.33	58.15	20.44	75.15	0.82
3	123	15.70	0.28	0.56	0.50	0.08	0.28	57.09	21.36	77.02	0.13
4	1,149	15.70	0.29	0.54	0.50	0.13	0.33	60.11	21.27	81.11	2.37
5	118	15.70	0.28	0.54	0.50	0.31	0.46	71.60	20.68	89.00	0.00
6	890	15.71	0.29	0.47	0.50	0.14	0.35	17.80	3.33	36.07	16.54
7	7,412	15.72	0.29	0.52	0.50	0.13	0.34	36.89	12.37	39.41	16.58
8	506	15.72	0.28	0.00	0.00	0.11	0.31	72.57	5.31	27.38	5.72
9	1,134	15.72	0.29	0.00	0.00	0.12	0.32	73.38	5.76	61.90	8.60
10	1,307	15.71	0.29	1.00	0.00	0.11	0.32	73.18	5.32	43.32	16.24
11	287	15.70	0.29	1.00	0.00	0.12	0.33	75.34	6.88	70.13	1.38
Total	13,836	15.72	0.29	0.51	0.50	0.13	0.33	50.08	22.42	49.91	22.18

Note: Groups (terminal nodes) are arranged level by level from left to right (terminal nodes begin to occur at the third level).
SES = socioeconomic status. Means for Male and Immigrant represent proportions of male students and immigrant students.

calculated as 0.80 divided by the square root of 92 (i.e., 0.08), and t = (13.67 – 49.91)/0.08 = –453.00 (df = 91). Therefore, mother SES is one critical independent variable that makes the first group depart from the population.

The idea of profits as discussed earlier adds interesting information to the interpretation of each terminal node (group) and the comparison across the groups. Table 5.7 presents the estimates of the profits for all of the 11 groups. In this case, taking any one course has a return (i.e., revenue) of 2 and an effort (i.e., expense) of 1, and each group has a profit value (revenue minus expense) that shares the same unit. All groups show profits. For example, the first group with 92 tenth graders profits the least as a whole group with a profit value of 0.772. The fifth group with 118 tenth graders profits the most as a whole group with a profit value of 1.864, more than double the profit value of the first group. Other groups have a profit value in between. A total of four groups manage to gain profits that double the profit value of the first group. There is a useful way to think of profits in this case. Given that the return of taking any one course is 2, the profit value for, say, the fifth group with 118 tenth grades (1.864) is equivalent to what would come from taking almost another course. In other words, the way that the tenth graders behave in this group is as if they had taken an extra course on top of the coursework they have taken in reality. Hopefully, interpretations like this may inspire one to think of innovative ways to attach meanings to the idea of profits.

Because costs and priors share a similar function to control misclassification (only from different perspectives), a CT analysis incorporating

TABLE 5.7 Estimates of Coursework Profits for Terminal Nodes (Groups)

Group	*N*	Percentage	Profit
1	92	0.66	0.772
2	818	5.91	1.587
3	123	0.89	1.260
4	1,149	8.30	1.687
5	118	0.85	1.864
6	890	6.43	1.192
7	7,412	53.57	1.367
8	506	3.66	1.534
9	1,134	8.20	1.433
10	1307	9.45	1.611
11	287	2.07	1.467

priors is omitted here except for the SPSS Decision Tree syntax in Appendix F (see the PRIORS CUSTOM subcommand). The pattern of "1 [1] 2 [2] 3 [2] 4 [3] " means that the prior probability of taking 2 = physics doubles [2] that of taking 1 = none [1], the prior probability of taking 3 = chemistry doubles [2] that of taking 1 = none [1], and the prior probability of taking 4 = both triples [3] that of taking 1 = none [1]. Finally, the three unique functions of costs, priors, and profits can be used either individually or collectively as long as the application can be justified.[4]

The three applications of CART in this chapter illustrate the point that not only is CART an effective analytical tool by itself (this point has also been shown in the previous chapters), but also it can effectively participate in the traditional type of data analysis with a great potential to enhance certain components of traditional data analysis or to create favorable conditions to improve the results of traditional data analysis. The analytical power of CART can be excellently extended and appreciated in combination with other statistical methods. This point is further illustrated in much detail in the following chapter.

Notes

1. In carrying out these applications, an effort is made to review and apply various CART techniques (e.g., using costs and priors) discussed in the previous chapters that can be manipulated in the software program of SPSS Decision Tree. The purpose is to demonstrate how to apply these techniques in the estimation or production of a CART tree.
2. This figure is the same as Figure 4.1 in the previous chapter and is reproduced here for the purpose of easier references when interpreting the results.
3. This table represents multiple logistic regression analyses with the cognitive and affective impact factors based on the 10 strata of data obtained in the CART analysis. Because CART is the main theme of this book, the second step of the analysis in Zhang and Bracken (1996) is simplified here for a better focus on the CART tree (the first step of the analysis).
4. The use of costs and priors does not entirely address the issue of misclassification. In fact, they work under some assumptions of their own. Not all CT analyses need to use costs or priors for control of misclassification. For a CT analysis, when neither costs nor priors are specified, the risk estimate that the SPSS Decision Tree routinely produces is the expected error rate (i.e., the expected probability of making an error in classification using the model). When costs are specified, the risk estimate is no longer a probability but the expected costs of errors using the model. When priors are specified, the risk estimate is the expected error rate for a population with the same distribution of priors across the categories of a dependent variable as one has specified. These statements are a direct interpretation of Table 3.4 obviously.

6

Advanced Techniques of CART

Although analytical strategies and practices are discussed and emphasized in this chapter on the advanced applications of CART, the premise of this chapter is that the pursuit of advanced techniques of CART would eventually come back to the common CART techniques that have already been discussed in the previous chapters. Without any new CART concepts or procedures in this chapter, one can concentrate on the "marriage" of CART with other statistical techniques for the enhancement of the analytical capacity of CART.

Extending Analytical Power of CART

A point made and emphasized in the previous chapter is that a combination of CART with other statistical procedures, either traditional or contemporary, can greatly extend the analytical power of CART. One can, to some extent, sense this point from the first application in the previous chapter in which CART and HLM work together for a complete data analysis of the rate of growth in mathematics achievement during the middle and high

Using Classification and Regression Trees, pages 101–128

school years. The joint of CART with HLM creates what can be referred to as a "hybrid" statistical model. In fact, hybrid statistical models are an easy and powerful way to extend the analytical power of CART (see more discussion later on). In this chapter, efforts are sought to create hybrid CART models for the extension of the analytical power of CART. These efforts pertain to the "between method" extension.

Another way to extend the analytical power of CART pertains to the "within method" extension; that is, one seeks efforts within the analytical category of CART that overcome some limitations of CART or enhance some functions of CART. This section focuses on these efforts, with the introduction of CHAID (chi-square automatic interaction detector), while leaving the efforts for the between method extension to specific sections to come in this chapter. CHAID was developed by Gordon Kass in 1980. Like CART, CHAID aims to reveal in a tree format the complex relationships among the independent variables that channel cases into different terminal nodes to account for the variation in a dependent variable. Also like CART, CHAID can build trees for nominal, ordinal, and continuous data. Given the analytical goal and purpose, CART and CHAID belong very much to the same family of analytical techniques.[1]

CHAID is considered in this book an extension of the analytical power of CART, simply because CHAID allows one to have multiple splits of a parent node into child nodes. CART, on the other hand, performs only a binary split of a parent node into two child nodes. For these reasons, some researchers consider CHAID producing "bushes" as opposed to CART producing trees. Such an extension obviously reveals more complex relationships among the independent variables, for which some of the hybrid statistical models to be discussed later on seek. One other sometimes desirable characteristic of CHAID is that when it splits a parent node according to a continuous independent variable it does so to create child nodes with approximately equal number of cases. This characteristic is also desirable for making policy and practice implications, because it avoids extreme terminal nodes for more compatible implications for policy and practice.

Fortunately, most statistical software programs for CART analysis include CHAID as an option of tree growth. For example, the SPSS Decision Tree software program which is applied to data analysis in this book offers CHAID. The SPSS output for a CHAID analysis is almost identical in format to the SPSS output for a CART analysis. Furthermore, the specification and interpretation of a CHAID tree are also very similar to the specification and interpretation of a CART tree. These functional similarities between CART and CHAID effectively avoid treating CHAID as a brand new statistical technique. Again, all of the differences between CART and CHAID in terms of functionality can be summarized as binary splits for CART versus multiple splits for CHAID. A CHAID analysis is upcoming in one of the sections to follow.

Concept of Hybrid Statistical Models

As defined earlier, a hybrid statistical model joins two or more statistical models or techniques together for an integrated data analysis where each statistical model has its own unique functions to link with other statistical models. For example, the first application of CART in the previous chapter joins CART and HLM together for a complete data analysis of the rate of growth in mathematics achievement during the middle and high school years, with the use of HLM to produce the dependent variable and the use of CART to explore the complex relationship between the dependent variable (newly created by HLM) and the independent variables.

The hybrid statistical models do not necessarily imply the creation of a new statistical theory; rather, they are a creative or innovative application of some existing statistical theories. The hybrid statistical models simply create an analytical framework or environment where one can take advantage of the strengths of more than one statistical model. In other words, they are geared more towards theory application rather than theory development. Nonetheless, as just argued, the hybrid statistical models do create new and innovative analytical frameworks or environments for more effective and efficient data analysis. With the following sections, this chapter is heavily devoted to the development of the hybrid statistical models for the extension of the analytical power of CART.

Longitudinal CART Analysis

With longitudinal data, especially panel data, one's analytical interest focuses usually on the change over time of cases (e.g., students) concerning a trait or behavior. It is therefore logical to use the rate of change (either linear or nonlinear) concerning the trait or behavior as a "summarized" indicator of the trait or behavior of the cases. In this sense, the rate of change in the trait or behavior becomes the outcome measure or the dependent variable. This outcome measure or dependent variable offers itself for all kinds of data analysis including CART. The first application of CART demonstrated in the previous chapter is a good example of a longitudinal CART analysis. Again, 6 years of panel data are ideal for either a linear or a nonlinear specification of change in terms of mathematics achievement over the entire middle and high school years (Grades 7 to 12). One may recall that the application proceeds to create a hybrid statistical model for a complete data analysis of the rate of growth.

The hybridization of CART with HLM moves data analysis to a whole new level in terms of the application of CART. CART may not be an appropriate statistical technique to decide on the nature and complexity concerning the rate of change in mathematics achievement over the entire middle and high school years. HLM, on the other hand, possesses this capacity, because multiple HLM models with linear and nonlinear rates of growth can be specified as well as compared and contrasted for model-data-fit statistics to identify the most appropriate form or specification of change. Once HLM captures the best approximation of growth in mathematics achievement over the entire middle and high school years, CART effectively channels students into various categories of growth based on individual characteristics of students. This "marriage" between CART and HLM creates a new and innovative analytical framework or environment for longitudinal CART analysis. One has witnessed in the previous chapter that this analytical framework or environment is capable of generating results or findings that cannot be obtained through traditional statistical techniques such as multiple regression analysis.

This hybrid CART model can handle any number of time points. In the case of the pretest and posttest design, the gain (score) can be created as the primitive type of the rate of change. If one does not desire to utilize the concept of gain, one can use the posttest measure as the dependent variable. One can then "force" the pretest measure to be used as the first independent variable to partition the root node. Most statistical software programs for CART analysis such as the SPSS Decision Tree program do allow one to specify a forced selection of a certain independent variable as the first independent variable to partition the root node. This action effectively takes into account the interaction of the pretest measure with the posttest measure (or the impact of the pretest measure on the posttest measure). Longitudinal designs with three or more time points can fit directly into HLM for the specification of either linear or nonlinear change. Again, the first application of CART in the previous chapter is a good example for a longitudinal CART analysis with three or more time points. One can easily apply this hybrid CART model to data obtained from both simple and complex longitudinal designs.

Multivariate CART Analysis

Following the conventional notion of multivariate statistics, if one has two dependent variables that are correlated with each other, two univariate analyses examining each dependent variable separately are not appropriate. In this case, a multivariate technique is needed to analyze the two dependent variables together in one analysis. For example, when an experimental design (i.e., treatment group versus control group) produces two outcome measures that are correlated with each other, multivariate analysis of variance (i.e., MANOVA) instead of analysis of variance (ANOVA) should be performed. In sum, multivariate statistics are needed when dealing with more than one dependent variable that are correlated with one another.

In the application of CART, similar situations can also occur in which there are more than one dependent variable that are correlated with one another. Following the same logic, separate CART analyses examining each dependent variable separately are not appropriate. There is a need for a multivariate CART analysis. To address the lack of information in the literature on multivariate CART models, this book proposes an innovative analytical framework or environment for multivariate CART analysis.

First of all, one may desire to know how strongly the dependent variables must correlate with one another in order to "qualify" for a multivariate CART analysis. The same conventional criteria can apply here. According to Tabachnick and Fidell (2007), when the dependent variables are moderately correlated (i.e., $.40 \leq r \leq .70$), multivariate statistical techniques are needed. When the dependent variables are highly correlated (i.e., $r > .70$), one should consider data reduction techniques such as factor analysis to combine the dependent variables. When the dependent variables are weakly correlated (i.e., $r < .40$), one can consider univariate statistics (i.e., separate analyses of the dependent variables). Following the same recommendations, separate CART analyses can be performed when the dependent variables have $r < .40$, one CART analysis for each dependent variable. When the dependent variables have $.40 \leq r \leq .70$, multivariate CART analysis is called for. When the dependent variables have $r > .70$, factor analysis can be employed to, say, create one general factor, and in this case one CART analysis can be performed on this factor.

When a multivariate CART analysis is called for, one needs to examine the measurement scales of the moderately correlated dependent variables. For the purpose of demonstration, this chapter considers the case of two moderately correlated dependent variables. There are three possible combinations based on the measurement scales of these dependent variables (two dichotomous dependent variables, two continuous dependent variables, and one dichotomous dependent variable and one continuous dependent variable). This section addresses analytical strategies and practices separately for each combination.

Correlated Categorical Dependent Variables

In the case where the two (moderately correlated) dependent variables are categorical, the main strategy to develop a multivariate CART framework or environment is to combine the categorical dependent variables into a new single dependent variable so that the CART techniques that have been discussed in the previous chapters can be applied directly. For the illustrative purpose, one can think of two (correlated) dependent variables A and B with dichotomy (0, 1). So one has A(0, 1) and B(0, 1). These two dummy dependent variables can be combined into a new single dependent variable with four categories (Category 1 = A0 and B0, Category 2 = A0 and B1, Category 3 = A1 and B0, and Category 4 = A1 and B1). With the creation of this new categorical dependent variable (with four categories), the CART techniques can directly apply to the data. As a matter of fact, this strategy makes a multivariate CART analysis no different from a regular (univariate) CART analysis.

Consider one example where one intends to examine the individual characteristics of school-aged children that are associated with drinking and smoking. When drinking and smoking are correlated as outcome measures, the above strategy creates four categories: 1 = both (i.e., drink and smoke), 2 = drink only (i.e., drink but not smoke), 3 = smoke only (i.e., smoke but not drink), and 4 = neither (i.e., not drink and not smoke). Based on the individual characteristics, a CART analysis can channel children into the four categories with different predominance of categories in different terminal nodes. The data describe a Canadian national (representative) sample of 2,310 tenth grade students from the cross-national survey on Health Behaviors in School-Aged Children (HBSC). The four categories combining drinking and smoking (created above) form the dependent variable. The independent variables include mental health (condition; on a scale of 0–12 with a higher value indicating a worse condition); physical health (condition; on a scale of 0–20 with a higher value indicating a worse condition); (ease of) making friends (on a scale of 1–4 with a higher value indicating a better chance of making friends); (feeling) helpless (yes, no); and (worrying about) body imagine (yes, no).

The SPSS syntax to run the CART analysis is shown in Appendix G. In this particular analysis, pruning for the tree is specified (see the METHOD TYPE subcommand where PRUNE=SE(1) indicates pruning). The CART tree is specified to grow eight levels (resulting in a huge tree) before pruning takes effect (see the GROWTHLIMIT subcommand). The results of this multivariate CART analysis are presented in Figure 6.1. The final CART tree contains three levels. The root node (i.e., the national sample)

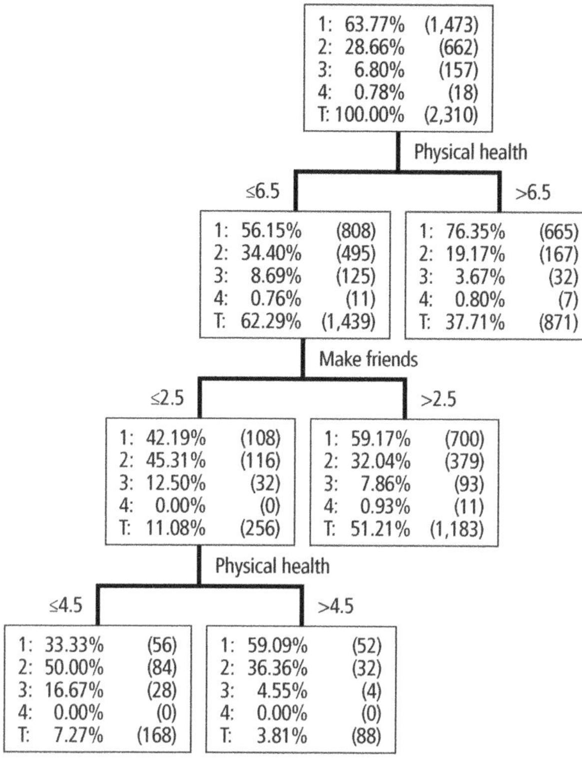

Figure 6.1 CART tree on drinking and smoking of tenth grade students. In each box, the first column indicates categories with T = total. The percentage and the number of students of each category follow.

is partitioned first by physical health (condition), resulting in one terminal node on the right. The parent node on the left is then partitioned by (ease of) making friends, resulting in one terminal node on the right. The parent node on the left is then partitioned by physical health (condition) again, resulting in two terminal nodes at the end or top of the tree. Physical health (condition) turns out to be the most important independent variable to the multivariate relationship between drinking and smoking among school-aged children (i.e., the tenth graders). In addition, (ease of) making friends is another important independent variable to the multivariate relationship. Other independent variables including mental health (condition), (feeling) helpless, and (worrying about) body imagine are not important to the multivariate relationship.

The interpretation of Figure 6.1 is centered around the multivariate relationship between drinking and smoking. Given that data analysis is performed on a national representative sample, the results from each terminal node (and the root node) are generalizable to the population. Thus, 63.77% of the population of the tenth graders (in Canada) both drink and smoke; meanwhile, 28.66% drink only (i.e., drink but not smoke), 6.80% smoke only (i.e., smoke but not drink), and between seven and eight out of one thousand refrain from both. Overall, from the drinking perspective, 63.77 + 28.66 = 92.43% of the population drink and from the smoking perspective, 63.77 + 6.80 = 70.57% of the population smoke.

The first partition by physical health (condition) produces the terminal node with the highest percentage of the tenth graders who both drink and smoke among all terminal nodes (and the root node). This terminal node represents more than a third of the population (i.e., the tenth graders; $N = 871$ which is 37.71% of the population). Among these tenth graders, 76.35% both drink and smoke. Interestingly (and to some extent surprisingly), it is the tenth graders with more physical health problems (i.e., worse physical health) who tend to both drink and smoke, given that this terminal node contains the tenth graders with physical health (condition) scores larger than 6.5 (on a measurement scale of 0–20). Apart from the above characteristics, from the drinking perspective, 76.35 + 19.17 = 95.52% of the tenth graders in this subpopulation drink; from the smoking perspective, 76.35 + 3.67 = 80.02% of the tenth graders in this subpopulation smoke; and eight out of one thousand in this subpopulation refrain from both. This terminal node is definitely the highlight of this multivariate CART analysis, identifying the most problematic subpopulation of the tenth graders compared with the general population (i.e., the root node).

The second partition by (ease of) making friends produces the majority terminal node ($N = 1,183$ which is 51.21% of the population). This majority subpopulation can be characterized as having few physical health problems [with physical health (condition) scores smaller than or equal to 6.5 on a measurement scale of 0–20] and making friends easily [with (ease of) making friends scores large than 2.5 on a measurement scale of 1–4]. In this majority subpopulation of the tenth graders, 59.17% both drink and smoke. From the drinking perspective, 59.17 + 32.04 = 91.21% of the tenth graders in this majority subpopulation drink; from the smoking perspective, 59.17 + 7.86 = 67.03% of the tenth graders in this majority subpopulation smoke; and between nine and ten out of one thousand in this majority subpopulation refrain from both. This majority subpopulation is very similar to the general population (i.e., the root node).

Although more than half of the majority subpopulation both drink and smoke, it has fewer tenth graders both drinking and smoking compared with the tenth graders in the problematic subpopulation, down 76.35 – 59.17 = 17.18

percentage points. In addition, the majority subpopulation has overall drinking down 95.52 – 91.21 = 4.31 percentage points even though drinking is an epidemic problem in the majority subpopulation, and meanwhile, overall smoking comes down 80.02 – 67.03 = 12.99 percentage points. Finally, the majority subpopulation shows slight improvement from the perspective of the tenth graders who refrain from both drinking and smoking.

The third partition by physical health (condition) again produces two minority terminal nodes. With $N = 168$ which is 7.27% of the population, the left one represents the physically healthiest tenth graders who do not make friends easily [physical health (condition) scores smaller than or equal to 4.5 on a measurement scale of 0–20 and (ease of) making friends scores smaller than or equal to 2.5 on a measurement scale of –4]. The percentage of the tenth graders who both drink and smoke is dramatically lower in this subpopulation than in any other subpopulation (and the root node). Although this subpopulation contains a larger percentage of the tenth graders who drink only or smoke only, the percentage of overall drinking or overall smoking shows the most positive information. Drinking only is the predominant category among these tenth graders. The problem with this minority subpopulation (and the other one as well) is that there are not refrainers. The other minority subpopulation ($N = 88$ which is 3.81% of the population) resembles the majority subpopulation, and therefore, it identifies a subpopulation among those tenth graders who have few physical health problems but do not make friends easily which is similar to the majority subpopulation. This is another "wonder" that CART can do but no traditional statistical techniques such as multiple regression analysis can easily do. Although a split produces two tree branches that are significantly different, terminal nodes from both sides can still be similar. This is partially because the same independent variables can take part in multiple splits during the construction of a CART tree. Meaningful implications for policy and practice can result from this unique function of CART.

Correlated Continuous Dependent Variables

In the case of (two) correlated dependent variables that are both continuous, the main strategy is to use one of them as the dependent variable for CART and force the other as the first independent variable to partition or split the root node in the CART tree. The goal is to make the two dependent variables interact to establish the multivariate relationship between them. Among the two dependent variables, the choice of which one to become the dependent variable for CART is open. One option is to select the one with a better univariate performance. For example, each of the two dependent variables is regressed on the same set of independent variables to be used later on in the CART analysis. The dependent variable with a larger R^2 (i.e., the proportion of variance explained by the model) is chosen as the dependent variable for CART. The other variable is then forced to carry out the first partition of the root node. Another option is to use each dependent variable in an alternate way as the dependent variable for CART (and then force the other dependent variable as the first independent variable to partition the root node). The results concerning the dependent variable with a better CART tree become the final results for interpretation. For example, one may interpret the results concerning the dependent variable with a larger R^2 for CART.

Consider one example where one intends to examine both mental health and physical health in relation to individual characteristics among the (2,310) tenth graders from the same database that describes a Canadian national (representative) sample obtained from the HBSC. The measurement scales are the same as earlier for mental health (condition; on a scale of 0–12 with a higher value indicating a worse condition) and physical health (condition; on a scale of 0–20 with a higher value indicating a worse condition). The correlation between mental health and physical health is .55, ideal for a multivariate analysis (see Tabachnick & Fidell, 2007). Because CHAID allows multiple splits of a parent node into child nodes, it captures the (multivariate) relationship between mental health and physical health more fully. The employment of CHAID is appropriate in this case of application.

To choose the dependent variable for (multivariate) CHAID, mental health and physical health are used in an alternate way as the dependent variable for CHAID, with the other dependent variable forced as the first independent variable to partition the root node. The independent variables describe individual characteristics of the tenth graders, including gender (male, female), age, father socioeconomic status (SES), mother SES, and the number of parents (guardians; on a measurement scale of 0–2). Both father SES and mother SES are standardized variables. When mental health is used as the dependent variable for CHAID with physical health forced as the first independent variable to partition the root node, $R^2 = 29.21\%$. When physical health is used as the dependent variable for CHAID with mental health forced as the first independent variable to partition the root node, $R^2 = 27.51\%$. One can choose mental health as the dependent variable for CHAID and physical health as the first independent variable to partition the root node. Appendix H presents the SPSS Decision Tree syntax for this CAHID analysis. There is in the syntax the specification of the split sample validation that involves a training sample and a test sample (often half and half of the total sample; see the subcommand of VALIDATION TYPE in Appendix H). The idea is to work with the training sample to develop the tree and then validate the tree with the test sample. Finally, pruning is not available as a function of tree growth under CHAID (in the SPSS Decision Tree program). Figure 6.2 presents the CHAID tree based on the test sample (from the split sample validation).

First of all, the test sample (from the split sample validation) contains 1,225 tenth graders as the root node. The average mental health (condition) is 4.13 on a measurement scale of 0–12, indicating that the population of the tenth graders is on average mentally healthy with few mental health issues (given that a higher value indicates a worse mental health condition). The (multivariate) connection of mental health with physical health is addressed by forcing physical health (condition) to be the first independent variable to partition the root node. This (first) forced split in the CHAID analysis results in six child nodes, five of which are terminal nodes. Again, CHAID allows multiple splits of a parent node into child nodes, as opposed to CART allowing only binary splits of a parent node into child nodes. Overall, the multivariate pattern is very clear, indicating that mental health issues or problems rise as physical health issues or problems rise. Stated differently, mental health and physical health issues or problems rise together. Specifically, with the same direction of measurement for physical health (condition; a higher value indicates a worse physical health condition), the first terminal node on the far left identifies a subpopulation of the tenth graders (11.60% of the population) with exceptionally good conditions of mental health and physical health. The tenth graders from this subpopulation "score" 2.07 on mental health (condition; on a measurement scale of 0–12) and 1 on physical health (condition; on a measurement scale of 0–20), indicating a superb lack of mental health and physical health issues or problems. This combination of *little concern* about mental health and little concern about physical health is independent of individual characteristics of the tenth graders (in terms of gender, age, father SES, mother SES, and number of parents or guardians), because no individual characteristics of the tenth graders are able to distinguish out any segments of this subpopulation.

One of the (statistically) significantly different subpopulations of the tenth graders is next door accounting for 18.90% of the population. The tenth graders from this subpopulation score 2.57 on mental health (condition) and either 2 or 3 on physical health (condition). This subpopulation thus indicates a considerable absence of mental health and physical health issues or problems. Similar to the previous terminal node, this combination of *minor concern* about mental health and minor concern about physical

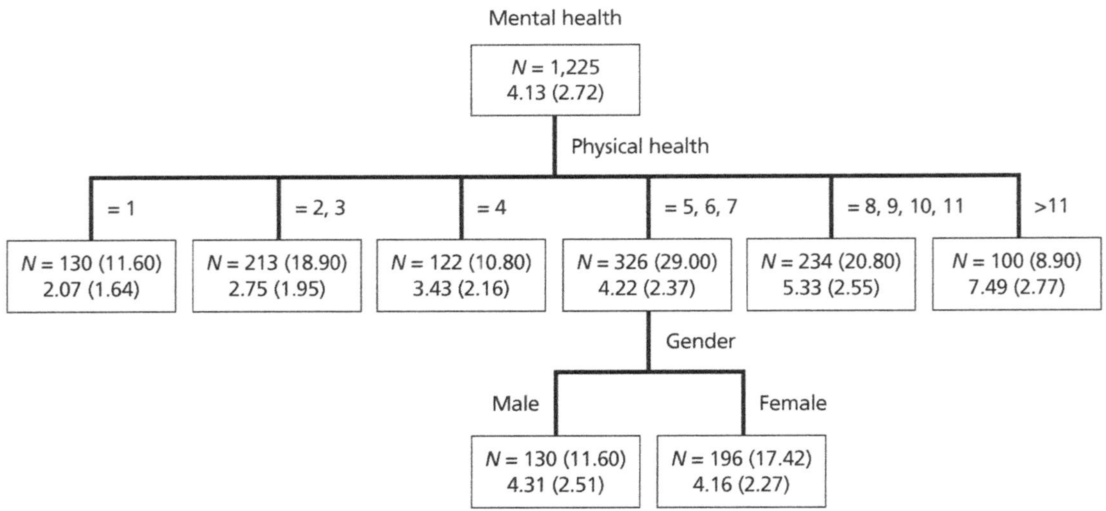

Figure 6.2 CHAID tree of multivariate relationship between mental health and physical health in relation to individual characteristics. In each node, top values indicate number of students with percentage of population in parenthesis, and bottom values indicate average physical health (condition) with standard deviation in parenthesis.

health is also independent of individual characteristics of the tenth graders. The same can be said with a little more compromise on mental health and physical health for the next terminal node representing a subpopulation which is 10.80% of the population.

The largest subpopulation which constitutes 29.00% of the population successfully represents the average. The tenth graders from this (popular) subpopulation score 4.22 on mental health (condition) and from 5 to 7 on physical health (condition). With between 4 and 5 out of 12 (i.e., a scale of 0–12) for mental health and from 5 to 7 out of 20 (i.e., a scale of 0–20) for physical health, this subpopulation rapidly approaches the averages of mental health and physical health issues or problems. Unique to this subpopulation, the gender of the tenth graders accounts for this (multivariate) combination of *moderate concern* about mental health and moderate concern about physical health, because gender is the most significant or important individual characteristic that partitions this popular subpopulation. With physical health (condition) remaining the same (i.e., from 5 to 7 in terms of scores), the male tenth graders are a smaller category within this popular subpopulation but with significantly more mental health issues or problems (i.e., showing a mental health condition score of 4.31), and the female tenth graders are a larger category within this popular subpopulation but with significantly less mental health issues or problems (i.e., showing a mental health condition score of 4.16). In fact, this popular or average subpopulation is the only one that has a significant or important relationship with individual characteristics of the tenth graders.

The deterioration in mental health and physical health starts in the terminal node to the right of the above popular subpopulation. This subpopulation constitutes 20.80% of the population. The tenth graders from this subpopulation score 5.33 on mental health (condition) and from 8 to 11 on physical health (condition). This subpopulation can be reasonably labeled as a subpopulation at risk. This combination of *high concern* about mental health and high concern about physical health is independent of individual characteristics of the tenth graders. The terminal node on the far right represents the problematic subpopulation (8.90% of the population). Although this subpopulation is the smallest among all subpopulations, the mental health and physical health conditions are worrisome with a score of 7.49 out of 12 (i.e., a scale of 0–12) on mental health (condition) and a score of 12 and above out of 20 (i.e., a scale of 0–20) on physical health (condition). This subpopulation may well be the highlight of the whole CHAID tree, suggesting that nearly one in ten in the population of the tenth graders are in need of interventions or treatments. This combination of *serious concern* about mental health and serious concern about physical health is independent of individual characteristics of the tenth graders.

Because this is a multivariate analysis, the scores on mental health (condition) and physical health (condition) in each subpopulation may be considered as "weights" for the construction of a linear composite of mental health (condition) and physical health (condition), very much similar to the concept of linear composites in canonical analysis. The term, combination, as applied many times above is purposefully chosen to imply the idea of composite. Interpretative languages such as a combination of moderate concern about mental health (condition) and moderate concern about physical health (condition) for the most popular subpopulation are a good way to "attach weights" to a subpopulation, highlighting the multivariate nature of mental health and physical health. When a parent node is partitioned into child nodes such as the splits in this particular CHAID application, it is actually this multivariate combination or relationship that is partitioned into a certain number of child nodes.

Correlated Dichotomous and Continuous Dependent Variables

When the (correlated) dependent variables are a combination of a categorical variable and a continuous variable, the analytical strategy is to employ a CHAID analysis where the categorical variable is the dependent variable and one forces the continuous variable to be the first independent variable to partition the root node. In this way, what has been discussed in the case of two continuous dependent variables can be carried out in terms of the specification of the CHAID model and the interpretation of the results. A less desirable strategy is to convert the continuous variable into a categorical variable so that what has been discussed in the case of two categorical dependent variables can be carried out in terms of the specification of the CART model and the interpretation of the results.

Multilevel CART Analysis

In this section, a brief discussion on the basic multilevel models is first provided to lay the background for multilevel CART analysis.[2] The main strategy for multilevel CART analysis is then introduced, with one example of application to follow.

Basics of Multilevel Models

Multilevel modeling has become the primary statistical technique to handle data with hierarchical or nested structures (e.g., individuals nested within groups). With individuals nested within groups, a multilevel model contains two levels; individuals are situated at the first level (i.e., the individual level) and groups are situated at the second level (i.e., the group level). Some prefer the terms of "within-group model" and "between-group model" to distinguish the two analytical units. The following model describes the individual or within-group model

$$Y_{ij} = \beta_{0j} + \sum_{p=1}^{n} \beta_{pj} X_{pij} + \varepsilon_{ij}$$

where Y_{ij} is the value of the dependent variable for individual i in group j, β_{0j} is the intercept representing the average measure of the dependent variable for group j with adjustment over the independent variables of X_{pij} ($p = 1, 2, \ldots n$), β_{pj} is the slope or regression coefficient of X_p for group j, and ε_{ij} is the error term unique to each individual. The intercept is usually treated as a random effect (with an error term) at the group level, and each slope can be treated either as a random effect or a fixed effect (without an error term) at the group level. The following models describe the group or between-group model

$$\beta_{0j} = \gamma_{00} + \sum_{q=1}^{m} \gamma_{0q} Z_{qj} + U_{0j}$$

$$\beta_{pj} = \gamma_{p0} + \sum_{q=1}^{m} \gamma_{pq} Z_{qj} + U_{pj}$$

where γ_{00} is the grand average measure of the dependent variable, γ_{p0} is the average slope of X_p, and U_{0j} and U_{pj} are the error terms each unique to each group. One of the essential functions of a multilevel model is to examine the effects of the variables at the group level, Z_{qj} ($q = 1, 2, \ldots m$), on the intercept (related to the outcome measure or dependent variable) and the slope of X_p (related to the effects of X_p on the outcome measure or dependent variable). The above group-level models treat the intercept as a random effect with U_{0j}, assuming that the intercept varies across groups. The above group-level models also treat the slope of X_p as a random effect with U_{pj}, assuming that the effects of X_p vary across groups. The slope of X_p

can also be treated as a fixed effect without U_{pj}, assuming that the effects of X_p do not vary across groups. In this case, there is usually no need to employ group-level variables to model the slope of X_p.

Strategy for Multilevel CART Analysis

The goal of a multilevel CART analysis where there is a data hierarchy with cases nested within groups is to adjust the effects of the independent variables at the case and group levels by the data segments (i.e., the terminal nodes) that a CART analysis establishes among the cases. This is a recognition that the hidden data segments representing a unique relationship of the dependent variable with the independent variables among the cases may influence the effects of the independent variables at the case and group levels. To some extent, the idea is that cases nested within segments may function as a "competing" data hierarchy to the existing data hierarchy of cases nested within groups. In other words, the hidden data segments may "distort" the estimation concerning the effects of the independent variables at the case and group levels. Therefore, a control over the (hidden) data segments among the cases (ignoring the groups to which they belong) allows for a better estimation of the effects of the independent variables at the case and group levels. This strategy makes much sense especially when there is a large number of independent variables at the case level. Traditionally, the interactions among these independent variables are usually ignored in the regression-based techniques such as multilevel modeling (the multilevel modeling technique cannot adequately decompose the interactions among the independent variables anyway). Instead of assuming the (fake) lack of interactions among the independent variables as one traditionally does in multilevel modeling, CART can help inform multilevel modeling with data segments created by the interactions among the independent variables at the case level.

From the perspective of data hierarchy, cases can be nested naturally within groups (in the original data); meanwhile, cases can also be nested within segments generated by a CART analysis of the dependent variable in relation to the independent variables (at the case level). In other words, the data segments establish another data hierarchy with cases nested within segments. Now there are two competing hierarchical structures in the data. One has cases nested within groups, and one has cases nested within segments. The two hierarchies together create a cross table or cross classification with one dimension as groups and one dimension as segments. Cases then come into cells of this table based on their memberships with the two dimensions (see Table 6.1). To some extent, a new data hierarchy

TABLE 6.1 An Example of Cross Classification of Groups and Segments

	Segment 1	Segment 2	Segment 3
School 1	XX	X	XXXXX
School 2	XXXXX	XX	XXXXX
School 3	X	XXXXXX	XXX
School 4	XXX	XXXXX	X
School 5	XXX	XX	XX

is established with cases nested within cells cross classified by groups and segments. The cross classified multilevel models can readily work with this unique data hierarchy (e.g., Goldstein, 1995; Raudenbush & Bryk, 2002).

From the analytical perspective, a multilevel CART analysis entails two main steps. The first step in the data analysis is to perform a CART analysis to identify the (hidden) data segments among the cases. This implies that the CART analysis is performed on the entire sample of the cases ignoring the groups to which they belong. In the second step of the data analysis, multilevel modeling is performed on the data with two competing hierarchies (i.e., cases nested within groups and cases nested within segments). Again, this multilevel modeling is often referred to as multilevel cross classification modeling, and some multilevel modeling software programs such as HLM and MLwiN can estimate this type of multilevel models (e.g., Charlton, Rasbash, Browne, Healy, & Cameron, 2017).

As an application, multilevel CART modeling is performed on a nationally representative sample of students in the United States ($N = 5{,}712$) from the 2015 Programme for International Student Assessment (PISA). The PISA 2015 focuses on science education with measures of science achievement.[3] With students nested within schools, the current analysis attempts to examine both individual differences in science achievement (at the student level) and contextual effects on science achievement (at the school level). At the student level, individual differences in science achievement concern students' age, gender, SES (measured in PISA 2015 as economic, social, and cultural status or ESCS), immigration status, and home language. At the school level, contextual effects on science achievement concern mainly school socioeconomic composition or school mean SES (i.e., school mean ESCS in the case of PISA 2015).

Following the two steps approach in multilevel CART modeling, a CART analysis is first performed to identify the (hidden) data segments among the cases (ignoring the groups to which they belong). The purpose and function of this CART analysis are quite similar to those in the CART analysis of dropping out of advanced mathematics (see Application 2 in Chapter 5). The goal is to identify critical data segments (or data strata in the case of dropping out of advanced mathematics). Because of this similarity and to save some space, the SPSS Decision Tree syntax is omitted here, just to mention the specification of tree growth up to four levels and cross validation (with defaults for all other aspects of the tree). The CART analysis reveals 13 data segments created by the interactions among the independent variables (see Figures 6.3 and 6.4). Out of the five independent variables at the student level, four (age, gender, ESCS, and home language) take part in the formation of the data segments.[4] This CART tree therefore establishes the new data hierarchy of cases or students nested within segments created by the interactions among the independent variables.

Now the two nesting structures, students nested within schools and students nested within segments, create a case of cross classification with one dimension as schools and one dimension as segments. The HLM software program is then employed to perform a multilevel cross classification analysis. Table 6.2 presents the results of this analysis. This table can also be referred to as the results of the multilevel CART analysis because of the involvement of CART in the overall analysis. Table 6.2 includes two sets of models for comparison. Model 1 is often referred to as the "null" model without any independent variables at any level. Multilevel models like this are often used to partition the variance in the dependent variable. There are two models under Model 1, one estimated with only one nesting structure of students nested within schools whereas the other estimated as the cross classified model accommodating both nesting structures (i.e., with the addition of the nesting structure of students nested within segments). Model 1 portrays a very interesting and critical picture of the influence of the hidden data segments on the distribution of the total variance (in science achievement). The variance attributable to schools is dramatically decreased from 1,863.92 to 993.76 when the hidden data segments are taken into account. In other words, the hidden data segments, if considered, would cut the variance attributable to schools by nearly half amount. In fact, there is more variance attributable to segments than to schools. On the other hand, because CART is performed among students, one may have the impression that the hidden data segments would take away a lot of variance from the student level. This is not necessarily true. In Table 6.2, the variance attributable to students is only slightly decreased from 7,684.64 to 7023.97 when the hidden data segments are taken into account. In fact, the variance component that is influenced more by the hidden data segments is what is attributable to schools.

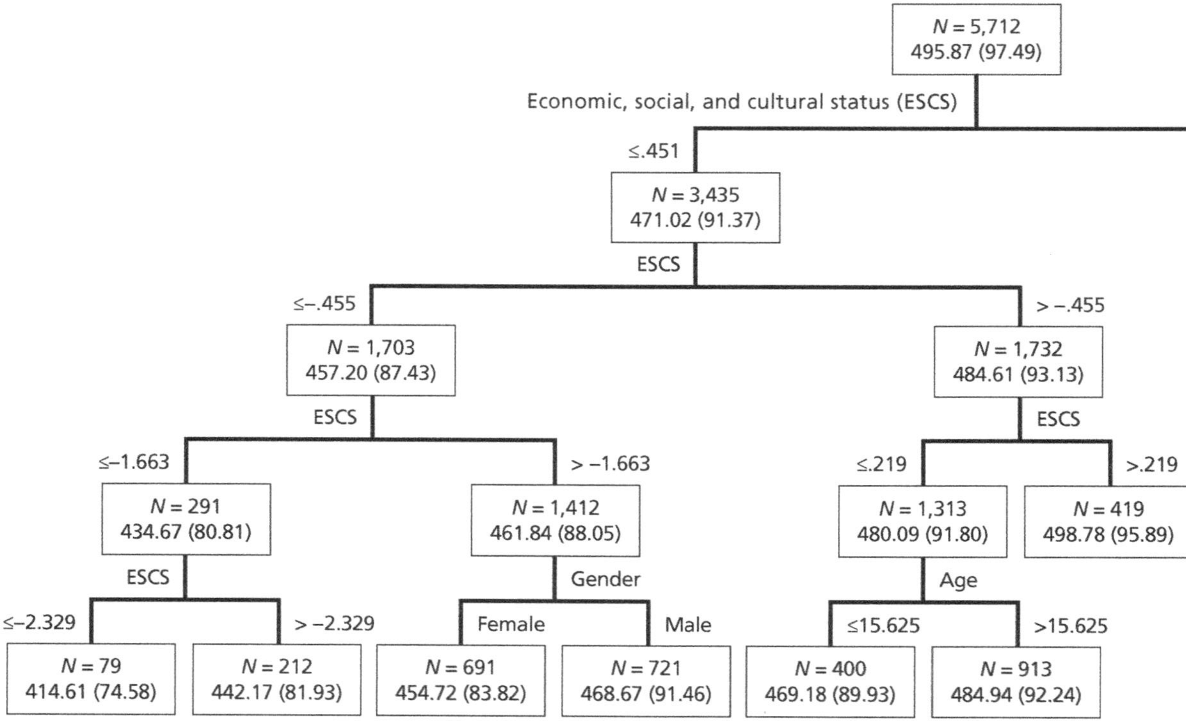

Figure 6.3 Partial (left) CART tree of science achievement conditional on student background variables. In each node, the top value indicates the number of students, and the bottom value indicates the average science achievement with standard deviation in parenthesis.

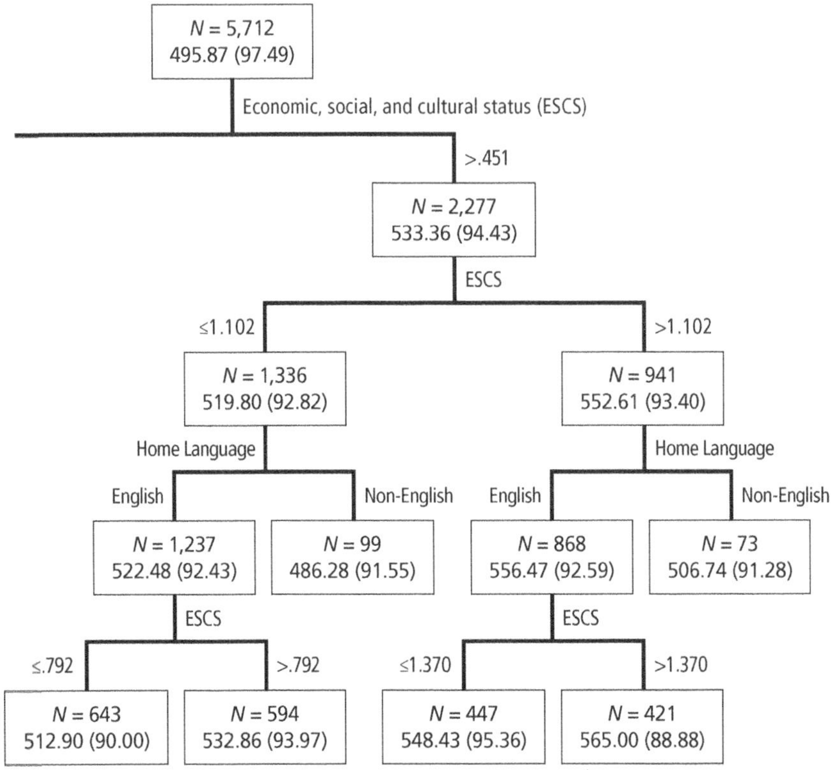

Figure 6.4 Partial (right) CART tree of science achievement conditional on student background variables. In each node, the top value indicates the number of students, and the bottom value indicates the average science achievement with standard deviation in parenthesis.

Meanwhile, Model 2 is often referred to as the "full" model with all independent variables at different levels. Model 2 portrays a very interesting and critical picture of the influence of the hidden data segments on the effects of the independent variables on science achievement at both student and school levels. The effects associated with age, ESCS, and home language on science achievement all decreased to a good degree. The most dramatic decrease concerns home language in that the statistically significant effects of home language actually disappear once the hidden data segments are taken into account. The decrease in terms of the effects of ESCS is also rather evident. Only do gender differences in science achievement remain relatively stable. Immigration status is "inactive" in both cases. Changes also occur at the school level. The current analysis includes only school mean ESCS at the school level. The effects of this school contextual variable on science achievement appears relatively stable.

TABLE 6.2 Nested Cross-Classified Multilevel Models Describing Relationship Between Science Achievement and Student-Level and School-Level Variables Taking Into Account Data Segments (Groups) Among Students Generated Through CART (N = 5,712)

	Model 1				Model 2			
	Schools only		Cross classification		Schools only		Cross classification	
Intercept	494.29*	3.48	494.71*	9.27	478.82*	5.60	482.37*	6.85
Student effects								
Age					12.51*	4.00	10.40*	4.46
Male					7.64*	2.30	8.12*	2.70
ESCS[a]					22.70*	1.57	10.78*	3.34
Immigration					7.12	4.68	7.15	4.18
Home language					15.09*	4.89	8.76	4.83
School effects								
Mean ESCS					34.24*	5.22	33.68*	4.69
Random effects								
Among groups			1,003.30				259.66	
Among schools	1,863.92		993.76		834.41		720.03	
Among students	7,684.64		7,023.97		7,206.37		7,012.43	

* $p < .05$
[a] ESCS = economic, social, and cultural status

In sum, this application of a multilevel CART analysis to the PISA data (i.e., the United States sample) illustrates to some extent the importance of taking the hidden segments of the data into consideration. The hidden data segments may substantially redistribute the variance in the dependent variable and may considerably influence the effects of the case-level and group-level independent variables on the dependent variable.

CART Procedure for Meta-Analysis

The logic to connect CART with meta-analysis is based on the unique function of CART to channel cases with similar behaviors or characteristics on the outcome measure of interest into distinct groups. The goal of meta-analysis is to synthesize empirical studies with various effect sizes on a certain outcome measure of interest. Empirical studies have different study features (e.g., sampling procedures, measurement properties, sample characteristics). If interactions among study features can channel effect sizes (from empirical studies) into various categories in a CART tree, then the CART tree by itself becomes a valuable part of the research synthesis in that the CART tree shows how study features work together to produce different categories of effect sizes. Again, the unique capacity of CART in decomposing interactions among study features over traditional statistical techniques puts CART in an advantageous position to discern how study features work together to produce different categories of empirical studies. Based on this research premise comes the attempt to develop the CART procedure for meta-analysis.

Actually, the CART procedure for meta-analysis is straightforward. In meta-analysis, each effect size is accompanied by a weight (often referred to as the "inverse variance weight"). Meta-analytic procedures adjust effect sizes by these weights in data analysis. In some CART software programs such as the SPSS Decision Tree, an analytical function is provided to take into account the special influence of a certain variable. This is where meta-analytic weights can be specified and, as a result, the CART tree is developed under the influence of the weights. This specification takes care of the pairing between effect size and weight, fulfilling the weighting requirement for meta-analysis. Study features then work together to (interactively) channel effect sizes into different categories under the influence of the weights.

As an application, the partial meta-analytic data with 80 effect sizes from Ma, Shen, Krenn, Hu, and Yuan (2016) are analyzed using the CART procedure for meta-analysis. Their meta-analysis examines the relationship between learning outcomes (from early childhood education) and parental involvement (in early childhood education of their children). The effect size statistic in this meta-analysis is obviously correlation (coefficient) between learning outcomes and parental involvement. The inverse variance weight is the sample size of each empirical study minus three (e.g., Lipsey & Wilson, 2001). Specifically for the partial meta-analytic data, reading (including language) ability is selected as the learning outcome. Study features include characteristics of parental involvement (D1: home discussion, D2: home supervision, D3: home-school connection, D4: volunteer work for school; yes vs. no) as well as characteristics of research design (experiment vs. either survey or observation), measurement (standardized measure vs. nonstandardized measure), and sample (minority sample vs non-minority sample).[5]

To illustrate the specification of the CART procedure for meta-analysis, the SPSS Decision Tree syntax is presented in Appendix I. The use of the meta-analytic weights is specified as the "influence variable" (see the INFLUENCE subcommand in which "w" is the name of the weight variable in the meta-analytic data). Given the presence of 80 effect sizes, the CART tree is specified to grow up to four levels, the minimum size for a terminal node is five effect sizes, and each parent node needs to have ten effect sizes (see the GROWTHLIMIT subcommand). The CART tree is also cross validated

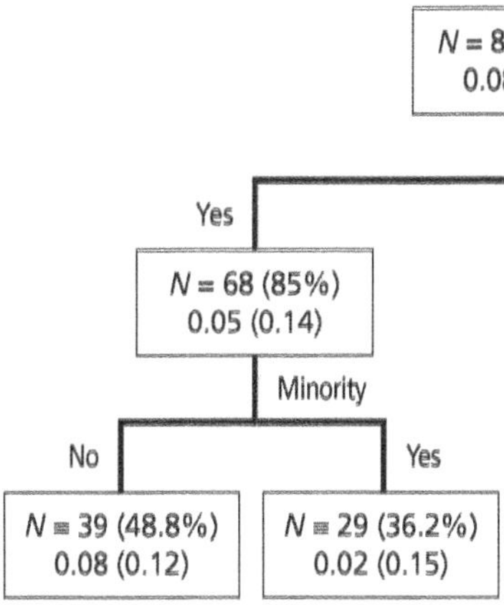

Figure 6.5 Results of CART procedure for meta-analysis of the relationship between reading outcome and parental involvement. In each node, the top value indicates the number of effect sizes with percentage in parenthesis, and the bottom value indicates the average effect size with standard deviation in parenthesis.

(see the VALIDATION TYPE subcommand). Figure 6.5 presents the CART tree that carries out the meta-analytic function.

In this figure, one substantive issue (related to parental involvement) and one methodological issue (related to sample characteristics) are the critical independent variables to partition effect sizes (representing individual empirical studies). Among these two issues, the substantive one is more important than the methodological one, as the first independent variable to partition the effect sizes. Of course, these issues work under the influence of the weights (i.e., the inverse variance weights). The root node has 80 effect sizes that represent 80 individual empirical studies. The average effect size for this meta-analysis is 0.08. Because correlation coefficient is employed as the effect size statistic for this meta-analysis, the correlation between reading outcome and parental involvement is 0.08, indicating a weak relationship.

The 80 individual empirical studies are partitioned according to one substantive issue of parental involvement, home supervision (D2), into two child nodes. One of the child nodes is a terminal node that gathers 12 individual empirical studies in which home supervision as a potential perspective of parental involvement is absent. These studies account for one sixth (i.e., 15%) of the individual empirical studies in the research literature. They indicate an average effect size of 0.26, more than three times stronger than the population average effect size of 0.08 in the root node. Therefore, these studies suggest a moderate relationship between reading outcome and parental involvement.

The other child node becomes a parent node of 68 individual empirical studies in which home supervision as a potential perspective of parental involvement is present. This parent node is further partitioned into two terminal nodes according to one methodological issue related to sample characteristics. With 39 individual empirical studies, the left terminal node accounts for nearly half (i.e., 48.8%) of the individual empirical studies in the research literature. These studies employ samples without minority children. They show a weak relationship between reading outcome and parental involvement, with an average effect size of 0.08 which is exactly the same as the population average effect size of 0.08 in the root node. Obviously, this terminal node contains the majority of the individual empirical studies in the research literature.

On the other hand, with 29 individual empirical studies, the right terminal node accounts for approximately one third (i.e., 36.2%) of the individual empirical studies in the research literature. These studies employ samples made of minority children. They show a lack of relationship between reading outcome and parental involvement, with an average effect size of 0.02, about four times weaker than the population average effect size of 0.08 in the root node. Altogether, an interesting pattern emerges concerning the distribution of the effect sizes across individual empirical studies in the research literature on the relationship between reading outcome and parental involvement. One third of the research literature indicates a lack of relationship; half of the research literature suggests a weak relationship; and the remaining one sixth of the research literature reveals a moderate relationship. The CART procedure for meta-analysis therefore identifies three categories of the research literature that are dramatically different in effect size. Table 6.3 shows the characteristics of these categories that help one understand the research literature in a unique way. Like all the cases earlier, this table contains descriptive information only and is thus easy to interpret.

For example, among the three categories of the research literature, the category with the strongest relationship shows that one in three individual empirical studies in this category has home discussion as one perspective of parental involvement, none of the individual empirical studies in this category has home supervision as one perspective of parental involvement, three in four individual empirical studies in this category have home-school connection as one perspective of parental involvement, and one in three individual empirical studies in this category has volunteer work for school as one perspective of parental involvement. In terms of study features, half

TABLE 6.3 Descriptive Information on Three Categories of Research Literature

	Category 1		Category 2		Category 3	
	Mean	SD	Mean	SD	Mean	SD
D1: home discussion	0.33	0.49	0.95	0.22	1.00	0.00
D2: home supervision	0.00	0.00	1.00	0.00	1.00	0.00
D3: home-school connection	0.75	0.45	0.85	0.37	1.00	0.00
D4: volunteer work for school	0.33	0.49	0.85	0.37	0.97	0.19
Research design	0.50	0.52	0.00	0.00	0.00	0.00
Measurement	1.00	0.00	1.00	0.00	1.00	0.00
Sample	0.25	0.45	0.00	0.00	1.00	0.00

Note: Categories are arranged according to the magnitude of the average effect size from large to small. The four dimensions of parental involvement are all coded as 1 = yes or present and 0 = no or absent. Research design is coded as 1 = experiment and 0 = either survey or observation. Measurement is coded as 1 = standardized measure and 0 = non-standardized measure. Sample is coded as 1 = sample made of minority children and 0 = sample without minority children. Mean for a variable indicates the percentage of individual empirical studies coded as 1 for the variable.

of the individual empirical studies in this category employ experiments to examine the relationship, all of the individual empirical studies in this category employ standardized measures to gauge reading outcome, and a quarter of the individual empirical studies in this category employ samples made of minority children. The other two categories of the research literature can be examined in the same way.

The partition by home supervision (D2) does not necessarily indicate that parental involvement programs without the perspective of home supervision promote higher reading outcome. Instead, these programs may very well not see any need for home supervision. On the other hand, when children's reading outcome is a concern (i.e., lower reading outcome), home supervision becomes one important way for parents to get involved for the purpose of improving reading achievement of their children. Overall, one can see that the CART procedure for meta-analysis synthesizes the research literature in an alternative manner, identifying categories of the research literature that are dramatically different in effect size.

Concluding Statement

In quantitative research, the univariate approach always concerns a dependent variable in relation to a set of independent variables. All statistical techniques under this approach (e.g., multiple regression analysis) find their unique ways to unearth this relationship. However, all these techniques assume a lack of interactions (or ignore the potential existence of interactions) among the independent variables during the modeling stage (i.e., when setting up models). Some researchers may include some interactive terms in their models, when there is a need to address part of their research questions. There is little purposeful intention to unearth the complex interactive relations among the independent variables potentially existing in the data as they relate to the dependent variable. As a result, an assumption is added to data analysis that there is no meaningful interactions among the independent variables that could impact the estimation of the parameters in a model.

CART offers a unique alternative to research and prediction. The uniqueness resides on the capacity of CART to extensively decompose the interaction effects among the independent variables as predictors of the dependent variable. In doing so, CART produces groups or terminal nodes with homogeneity within groups and heterogeneity between groups concerning the dependent variable that are a result of the complex interaction effects among the independent variables. A tree is a natural way to illustrate these identified interaction effects as they relate to the dependent variable.

As a data mining technique, CART works more efficiently and effectively with larger databases. In terms of model specification, CART allows one to exercise a great degree of control over the tree growth for various purposes of research and prediction. Once the results come out, the CART tree is easy to interpret (i.e., the interaction effects that produce the CART tree are easy to identify), and meanwhile, the characteristics of each group or terminal node are easy to sum up. Because CART identifies groups or terminal nodes with dramatically different measures on the dependent variable, CART informs policies and practices in a way to which few traditional statistical techniques can parallel.

The unique way of CART to specify the tree growth changes some conventional statistical conceptions such as the statistical significance of an independent variable. It is not that CART is unable to identify statistically significant independent variables. CART simply does it in a different (unconventional) way. The significance or importance of independent variables in CART speaks to the collective role of each independent variable in unearthing the complex interactive relations among the independent variables as they channel cases into unique groups or terminal nodes. To some extent, if one agrees with the notion that it is really the interaction effects among the independent variables that produce the relationship between the dependent variable and the independent variables, then one may even want to avoid discussing individual independent variables for significance or importance because an individual independent variable needs to be a part of an interaction in a CART tree to make itself significant or important. This is perhaps why the attention for interpretation of the CART tree is always on the groups or terminal nodes (in that they are the products of the interactions).

One way to extend the analytical power of CART is to incorporate CART with other statistical techniques (i.e., the concept of a hybrid model involving CART). All statistical techniques, either contemporary or traditional, exist for their unique capacities or purposes. Any combination of them (i.e., a hybrid model) gains "the best of both worlds." Following this logic, various attempts have been made in this book to build hybrid models in which CART plays at least an equally important role for data analysis. The hybrid models introduced in this book are by no means exclusive; instead, they function as an inspiration for researchers to build more hybrid models involving CART with more efficiency and effectiveness. It has been the purpose of this book all along to build a CART foundation strong enough for raising to higher levels of sophistication in using CART as a powerful tool for empirical research.

Notes

1. Apart from the major difference between CHAID and CART that pertains to the number of splits from a parent node into child nodes, there are some technical and functional differences between the two. Technically, CHAID utilizes a single dataset to build the tree, whereas CART utilizes a training dataset to build the tree and a "preserving" dataset to prune the tree. Also, CHAID uses a chi-square test for independence to examine the dependence between an independent variable and the dependent variable (if they are independent, there is no tree growth), whereas CART examines the amount of homogeneity within a node to determine whether to stop the tree growth. Functionally, CHAID may be more useful for research (i.e., analysis), where CART may be more useful for prediction (i.e., forecast). Stated differently, if the purpose of data analysis is to describe and understand the relationship between the dependent variable and the independent variables, CHAID may be more appropriate. If the purpose of data analysis is to develop a mechanism that efficiently and precisely classifies (new) cases, CART may be more appropriate.
2. HLM and multilevel modeling are interchangeable terms of the same statistical technique. HLM is used in the previous chapters to maintain consistency with what is used in the referenced articles. Multilevel modeling is used in this chapter as a more comprehensible term.
3. Based on "matrix sampling" in which students work on different (short) test booklets in order to cut down the time required for testing, PISA creates plausible values as the measure of academic achievement. In PISA 2015, each student has 10 plausible values on science achievement. These plausible

values are not test scores and thus cannot be used as such even though each plausible value shares the same measurement scale as the final measure of science achievement. They need to be integrated properly into one measure of science achievement. Some statistical software programs such as HLM can properly combine the 10 plausible values into a measure of science achievement. Because this issue is not directly relevant to the purpose of the discussion on multilevel CART analysis, this extra step in the data analysis is omitted. Instead, the first plausible value for science achievement is taken to function as the measure of science achievement only for the illustrative purpose.

4. The fact that four out of five independent variables take part in the CART analysis is a good indication that there indeed exist some important interactions among the independent variables. The negligence of these interactions in any modeling of the effects of the independent variables may create bias on the estimation.
5. For the purpose of demonstration, not all study features examined in Ma et al. (2016) are employed. Thus, the results here are illustrative of mainly the CART procedure for meta-analysis and cannot be considered as a replication of or a departure from the results of a previous meta-analysis.

References

Chapter 1

Aneshensel, C. S. (2002). *Theory-based data analysis for the social sciences.* Thousand Oaks, CA: Pine Forge.

Appel, K., & Haken, W. (1977). The solution of the four-color map problem. *Scientific American, 237,* 108–121.

Billey, S. (2015). *Computer assisted proofs: Coming soon to a theorem near you.* Retrieved from https://www.math.washington.edu/~billey/talks/MathDay.Computer.Proofs.pdf

Breiman, L. (2002, July). *The WALD Lecture II: Looking inside the black box.* Lecture featured at the 277th meeting of the Institute of Mathematical Statistics. Banff, AB.

Breiman, L., Friedman, J. H., Olshen, R. A., & Stone, C. J. (1984). *Classification and regression trees.* Belmont, CA: Wadsworth.

Clarke, A. E., Bloch, D. A., Danoff, D. S., & Esdaile, J. M. (1994). Decreasing costs and improving outcomes in systemic lupus erythematosus: Using regression trees to develop health policy. *Journal of Rheumatology, 21*(12), 2246–2253.

Cohen, J. (1988). *Statistical power analysis for the behavioral sciences* (2nd ed.). Hillsdale, NJ: Lawrence Erlbaum.

Data mining. (n.d.). In Merriam-Webster's online dictionary (11th ed). Retrieved from https://www.merriam-webster.com/dictionary/data%20mining

Efron, B. R., & Tibshirani, R. (1991). Statistical data analysis in the computer age. *Science, 253*(5018), 390–395.

Using Classification and Regression Trees, pages 129–132

Hoaglin, D. C., Mosteller, F., & Tukey, J. W. (Eds.) (2000). *Understanding robust and exploratory data analysis.* New York, NY: Wiley.

Hughes, K. D. (1999). *Gender and self-employment in Canada: Assessing trends and policy implications.* Ottawa, ON: Renouf.

Human Resources Development Canada & Statistics Canada. (2000). *Survey of self-employment: User's manual.* Available at http://www23.statcan.gc.ca/imdb/p2SV.pl?Function=getSurvey&SDDS=3850

Ma, X. (2003). Training behaviors of the self-employed in Canada: A decision tree analysis. In S. P. Shohov (Ed.), *Advances in psychology research* (Vol. 22, pp. 75–96). Hauppauge, NY: Nova Science.

McNamara, J. F. (2000). Teaching statistics in principal preparation programs. *International Journal of Educational Reform, 9*(4), 373–384.

Murray, S., & Zeesman, A. (2001). Introduction. In Statistics Canada & Human Resources Development Canada (Eds.), *A report on adult education and training in Canada: Learning a living* (pp. 5–10). Ottawa, ON: Authors.

Sinacore, J. M., Chang, R. W., & Falconer, J. (1992). Seeing the forest despite the trees: The benefit of exploratory data analysis to program evaluation research. *Evaluation and the Health Professions, 15*(2), *131–46.*

Srivastava, T. (2013, October 21). *Trick to enhance power of regression model.* Retrieved from https://www.analyticsvidhya.com/blog/2013/10/trick-enhance-power-regression-model-2/

Suyemoto, K. L., & MacDonald, M. L. (1996). The content and function of religious and spiritual beliefs. *Counseling and Values, 40*(2), *143–158.*

Tukey, J. W. (1977). *Exploratory data analysis.* Reading, MA: Addison-Wesley.

Ture, M., Kurt, I., Kurum, A. T., & Ozdamar, K. (2005). Comparing classification techniques for predicting essential hypertension. *Expert Systems with Applications, 29,* 583–588.

Zhang, H., & Singer, B. (1999). *Recursive partitioning in the health sciences.* New York, NY: Springer-Verlag.

Chapter 2

Breiman, L., Friedman, J. H., Olshen, R. A., & Stone, C. J. (1984). *Classification and regression trees.* Belmont, CA: Wadsworth.

Chebrolu, S., Abraham, A., & Thomas, J. P. (2005). Feature deduction and ensemble design of intrusion detection systems. *Computers & Security, 24*(4), 295–307.

Currie, C., Hurrelmann, K., Settertobulte, W., Smith, R., & Todd, J. (Eds.). (2000). *Health and health behaviour among young people* (Health Policy for Children and Adolescents, No. 1). Copenhagen, Demark: WHO Regional Office for Europe.

Diercks, D. B., Fonarow, G. C., Kirk, J. D., Emerman, C. L., Hollander, J. E., Weber, J. E., . . . ADHERE Scientific Advisory Committee and Investigators. (2008). Risk stratification in women enrolled in the Acute Decompensated

Heart Failure National Registry Emergency Module (ADHERE-EM). *Academic Emergency Medicine, 15*(2), 151–158.

Duda, R. O., Hart, P. E., & Stork, D. G. (2001). *Pattern classification* (2nd ed.). New York, NY: Wiley.

Lewis, R. J. (2000). *An introduction to classification and regression tree (CART) analysis.* Paper presented at the annual meeting of the Society for Academic Emergency Medicine. San Francisco, CA.

Morgan, J. N., & Sonquist, J. A. (1963). Problems in the analysis of survey data and a proposal. *Journal of American Statistical Association, 58*(302), 415–435.

Morrison, J. (1998). Introducing C.A.R.T. to the forecasting process. *Journal of Business Forecasting Methods & Systems, 17*(1), 9–12.

Morwitz, V. G., & Schmittlein, D. (1992). Using segmentation to improve sales forecasts based on purchase intent: Which "intenders" actually buy? *Journal of Marketing Research, 29,* 391–405.

Nelson, L. (1998). Recursive partitioning for the identification of disease risk subgroups: A case-control study of subarachnoid hemorrhage. *Journal of Clinical Epidemiology, 51*(3), 199–209.

Proriol, J. (1994). *Selection of variables for neural network analysis: Comparison of several methods with high energy physics data.* Aubiere Cedex, France: Laboratoire de Physique Corpusculaire de Clermont-Ferrand, Université Blaise Pascal.

Wang, M., Au, K., Ailamaki, A., Brockwell, A., Faloutsos, C., & Ganger, G. R. (2004). Storage device performance prediction with CART models. *Performance Evaluation Review, 32,* 412–413.

Chapter 3

Breiman, L., Friedman, J. H., Olshen, R. A., & Stone, C. J. (1984). *Classification and regression trees.* Belmont, CA: Wadsworth.

SPSS. (1999). *Introduction to AnswerTree.* Chicago, IL: SPSS Inc.

Tabachnick, B. G., & Fidell, L. S. (2007). *Using multivariate statistics* (5th ed.). Needham Heights, MA: Allyn & Bacon.

Zhang, H., & Singer, B. (1999). *Recursive partitioning in the health sciences.* New York, NY: Springer-Verlag.

Chapter 4

Breiman, L., Friedman, J. H., Olshen, R. A., & Stone, C. J. (1984). *Classification and regression trees.* Belmont, CA: Wadsworth.

Ma, X. (2005). Growth in mathematics achievement during middle and high school: Analysis with classification and regression trees. *Journal of Educational Research, 99*(2), 78–86.

Morrison, J. (1998). Introducing C.A.R.T. to the forecasting process. *Journal of Business Forecasting Methods & Systems, 17*(1), 9–12.

Ture, M., Kurt, I., Kurum, A. T., & Ozdamar, K. (2005). Comparing classification techniques for predicting essential hypertension. *Expert Systems with Applications, 29*(3), 583–588.

Chapter 5

American Association of State Colleges and Universities. (2006). High school coursework: Policy trends and implications for higher education. *Policy Matters, 3*(7).

Ma, X. (2005). Growth in mathematics achievement during middle and high school: Analysis with classification and regression trees. *Journal of Educational Research, 99*(2), 78–86.

Miller, J. D., Kimmel, L., Hoffer, T. B., & Nelson, C. (2000). *Longitudinal Study of American Youth: User's manual.* Chicago, IL: International Center for the Advancement of Scientific Literacy, Northwestern University.

National Council of Teachers of Mathematics. (2000). *Principles and standards for school mathematics.* Reston, VA: Author.

Raudenbush, S. W., & Bryk, A. S. (2002). *Hierarchical linear models* (2nd ed.). Newbury Parks, CA: SAGE.

Willett, J. B. (1988). Questions and answers in the measurement of change. In E. Z. Rothkopf (Ed.), *Review of research in education* (pp. 345–422). Washington, DC: American Educational Research Association.

Zhang, H., & Bracken, M. B. (1996). Tree-based, two-stage risk factor analysis for spontaneous abortion. *American Journal of Epidemiology, 144*(10), 989–996.

Chapter 6

Charlton, C., Rasbash, J., Browne, W. J., Healy, M., & Cameron, B. (2017). *MLwiN version 3.00.* Centre for Multilevel Modelling, University of Bristol.

Goldstein, H. (1995). *Multilevel statistical models* (2nd ed.). London, England: Edward Arnold.

Lipsey, M. W., & Wilson, D. B. (2001). *Practical meta-analysis.* Thousand Oaks, CA: SAGE.

Ma, X., Shen, J., Krenn, H. Y., Hu, S., & Yuan, J. (2016). A meta-analysis of the relationship between learning outcomes and parental involvement during early childhood education and early elementary education. *Educational Psychology Review, 28*(4), 771–801.

Raudenbush, S. W., & Bryk, A. S. (2002). *Hierarchical linear models* (2nd ed.). Newbury Parks, CA: SAGE.

Tabachnick, B. G., & Fidell, L. S. (2007). *Using multivariate statistics* (5th ed.). Needham Heights, MA: Allyn & Bacon.

APPENDIX A

Functionally Equivalent Binary Tree

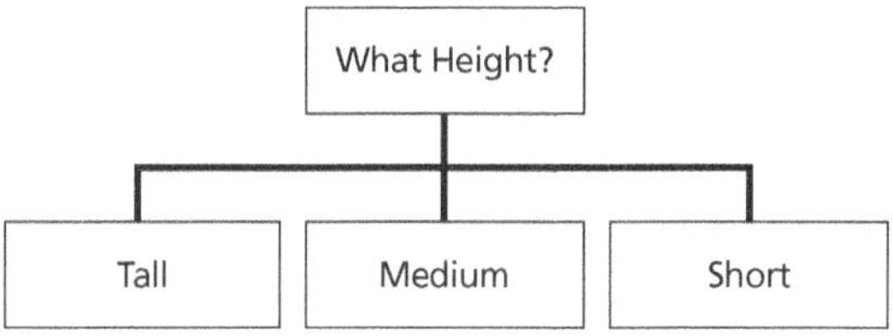

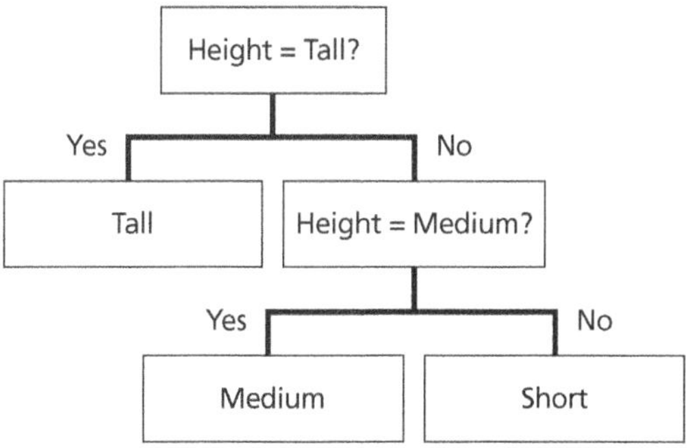

APPENDIX B

Common CART Software Programs

There are a few "off the shelf" statistical software programs or packages that are designed to at least partially facilitate a CART analysis. These programs share many common functions but differ in certain specific details (which in fact make some programs particularly powerful for certain types of CART analysis). The software programs that are most frequently used are introduced in this appendix. The introduction on each program focuses on its unique analytical functions.

C5.0

Often referred to as "statistical classifier," C5.0 (for Linux) and See5 (for Windows) are developed by Ross Quinlan for large scale data mining efforts to reveal patterns that delineate categories, assemble categories into classifiers, and use classifiers to make predictions. The program features the ease of use without any need for special statistical knowledge, the fast speed for data analysis, the maximum interpretability with classifiers expressed in the form of decision trees or if then rules, and the direct export

Using Classification and Regression Trees, pages 135–137

of classifiers into other computing systems. For further information, see http://www.rulequest.com/see5-info.html

CART

This software program features reliable pruning strategies, powerful binary split search approach, and automatic self-validation procedures. CART uses surrogate splitters to handle intelligently missing values, adjustable misclassification penalties to help avoid the most costly errors, and alternative splitting criteria to make progress when other splitting criteria fail. This program can import SAS, SPSS, Excel, and Lotus data files. For further information, see https://www.salford-systems.com/products/cart

DTREG

As a predictive modeling software, one of the main functions of DTREG is to compute and generate CART trees. This software program can be used to perform both CT (the dependent variable is categorical) and RT (the dependent variable is continuous). It has specific features to help one determine optimal tree size and apply variable costs and priors as well as variable weights in data analysis. For further information, see https://www.dtreg.com/

Precision Tree

Building decision trees and influence diagrams within Microsoft Excel, Precision Tree (as an add-on module to Excel) is fully integrated with spreadsheet models to allow one to visually map, organize, and analyze decisions. It is effective in dealing with complex and sequential (multistage) decisions, producing diagrams with nodes and branches to demonstrate different decision paths and chance events. Graphs and reports are customized using standard Excel features. Influence diagrams show results without being converted to decision trees. Results of decision analysis are updated automatically as models are changed. This program performs sensitivity analyses on any value in a decision tree or influence diagram. It cooperates with @RISK for complete Monte Carlo simulations. Trees and diagrams are easy to generate and edit. Probabilities are automatically normalized in chance nodes. Influence diagrams can describe asymmetric trees. Advanced features include logic nodes, reference nodes, custom VBA (visual basic for applications) utility functions, and linked trees. For further information, see http://www.palisade.com/precisiontree/

SPSS Decision Tree

This program is a part of SPSS, featuring scalable decision trees that reveal segments and predict group reactions (responses) to interventions (e.g., promotions). It has four algorithms for tree-based analysis, including CHAID (chi-square automatic interaction detection), exhaustive CHAID, CRT (i.e., CART), and QUEST (quick, unbiased, efficient statistical tree). At each step of tree growth, CHAID uses the independent variable with the strongest interaction with the dependent variable, while exhaustive CHAID examines all possible splits for each independent variable. QUEST is for nominal dependent variables only but runs fast and against the tendency of using independent variables with a large number of categories often occurred with other algorithms. This program displays models and results in various visual formats to help identify groups that matter the most. As part of SPSS, this program can import data files from many sources such as Excel, SAS, and ASCII. For further information, see https://www.ibm.com/ms-en/marketplace/spss-decision-trees

Selection of Programs

Learners may start to work with CART using SPSS Decision Tree to take advantage of its simplicity. This is a user-friendly program with enough functions to meet the vast majority of the analytical needs of the learners. DTREG is very much in the same category. For more advanced applications, C5.0 and CART are excellent choices. These are powerful programs designed mostly for those learning to become advanced or professional analysts. Precision Tree is not exactly CART, but it has some features that can be used to enhance certain perspectives of CART.

APPENDIX C

SPSS Decision Tree Syntax

This appendix presents the SPSS Decision Tree syntax that is used to perform the first application of CART in Chapter 5. This (set of) syntax can be directly copied and modified to perform a CART analysis. See the SPSS Decision Tree manual for more information: ftp://public.dhe.ibm.com/software/analytics/spss/documentation/statistics/24.0/en/client/Manuals/IBM_SPSS_Decision_Trees.pdf

```
TREE growth [s] BY gender [n] race [n] mothses [s]
fathses [s] age [s] siblings [s]
  /TREE DISPLAY=TOPDOWN NODES=STATISTICS
   BRANCHSTATISTICS=YES NODEDEFS=YES SCALE=AUTO
  /PRINT MODELSUMMARY RISK
  /GAIN SUMMARYTABLE=YES TYPE=[NODE] SORT=DESCENDING
   CUMULATIVE=NO
  /METHOD TYPE=CRT MAXSURROGATES=AUTO PRUNE=NONE
  /GROWTHLIMIT MAXDEPTH=4 MINPARENTSIZE=100 MINCHILDSIZE=50
  /VALIDATION TYPE=CROSSVALIDATION(10) OUTPUT=BOTHSAMPLES
  /CRT MINIMPROVEMENT=0.0001
  /MISSING NOMINALMISSING=MISSING.
```

Using Classification and Regression Trees, page 139

APPENDIX D

SPSS Decision Tree Output

This appendix represents partial SPSS output for the first application of CART. What is omitted in this appendix is the CART tree which is Figure 5.1. The three tables are mostly self-explanatory. The Model Summary table (Table D.1) indicates the selection of the variables, the specification of the model, and the structure of the tree. The Gain Summary for Nodes table (Table D.2) contains basic information on all terminal nodes with the Mean indicating (in this case) the (average) rate of growth in mathematics achievement during the entire middle and high school years for each terminal node. The Risk table (Table D.3) presents the risk estimate (1.482) for the calculation of the proportion of variance in the dependent variable accounted for by the CART tree.

Using Classification and Regression Trees, pages 141–142

TABLE D.1 Model Summary

Specifications	Growing Method	CRT
	Dependent Variable	growth
	Independent Variables	gender, race, mothses, fathses, age, siblings
	Validation	Cross Validation
	Maximum Tree Depth	5
	Minimum Cases in Parent Node	100
	Minimum Cases in Child Node	50
Results	Independent Variables Included	age, fathses, mothses, siblings, race, gender
	Number of Nodes	21
	Number of Terminal Nodes	11
	Depth	5

TABLE D.2 Gain Summary for Nodes

Node	*N*	Percent	Mean
20	64	2.1	4.2769
19	1,074	34.6	3.7835
13	312	10.1	3.7245
14	343	11.1	3.3611
16	60	1.9	3.3568
11	237	7.6	3.2131
4	407	13.1	3.1874
10	135	4.4	2.8711
12	321	10.3	2.7402
18	92	3.0	2.5585
17	57	1.8	2.1151

Growing Method: CRT
Dependent Variable: growth

TABLE D.3 Risk

Method	Estimate	Std. Error
Resubstitution	1.482	.044
Cross-Validation	1.528	.045

Growing Method: CRT
Dependent Variable: growth

APPENDIX E

SPSS Decision Tree Syntax Using Costs and Profits

```
TREE Course [n] BY age [s] gender [n] immig [n] sesm [s]
sesf [s]
  /TREE DISPLAY=TOPDOWN NODES=STATISTICS
   BRANCHSTATISTICS=YES NODEDEFS=YES SCALE=AUTO
  /DEPCATEGORIES USEVALUES=[1 2 3 4]
  /PRINT MODELSUMMARY CLASSIFICATION RISK
  /GAIN SUMMARYTABLE=YES TYPE=[NODE] SORT=DESCENDING
   CUMULATIVE=NO
  /METHOD TYPE=CRT MAXSURROGATES=AUTO PRUNE=NONE
  /GROWTHLIMIT MAXDEPTH=4 MINPARENTSIZE=100 MINCHILDSIZE=50
  /VALIDATION TYPE=NONE OUTPUT=BOTHSAMPLES
  /CRT IMPURITY=GINI MINIMPROVEMENT=0.0001
  /COSTS CUSTOM= 1 1 [0] 1 2 [1] 1 3 [1] 1 4 [2] 2 1 [1]
   2 2 [0] 2 3 [1] 2 4 [1] 3 1 [1] 3 2 [1] 3 3 [0] 3 4 [1]
   4 1 [2] 4 2 [1] 4 3 [1] 4 4 [0]
  /PRIORS FROMDATA ADJUST=NO
  /PROFITS CUSTOM=1 [0 0] 2 [2 1] 3 [2 1] 4 [4 2]
  /MISSING NOMINALMISSING=MISSING.
```

Using Classification and Regression Trees, page 143

APPENDIX F

SPSS Decision Tree Syntax Using Priors

```
TREE Course [n] BY age [s] gender [n] immig [n] sesm [s]
sesf [s]
  /TREE DISPLAY=TOPDOWN NODES=STATISTICS
   BRANCHSTATISTICS=YES NODEDEFS=YES SCALE=AUTO
  /DEPCATEGORIES USEVALUES=[1 2 3 4]
  /PRINT MODELSUMMARY CLASSIFICATION RISK
  /METHOD TYPE=CRT MAXSURROGATES=AUTO PRUNE=NONE
  /GROWTHLIMIT MAXDEPTH=4 MINPARENTSIZE=100 MINCHILDSIZE=50
  /VALIDATION TYPE=NONE OUTPUT=BOTHSAMPLES
  /CRT IMPURITY=GINI MINIMPROVEMENT=0.0001
  /COSTS EQUAL
  /PRIORS CUSTOM=1 [1] 2 [2] 3 [2] 4 [3] ADJUST=NO
  /MISSING NOMINALMISSING=MISSING.
```

Using Classification and Regression Trees, page 145

APPENDIX G

SPSS Decision Tree Syntax for Drinking and Smoking Data

```
TREE DS [n] BY mentheal [s] physheal [s] makefrie [s]
helpless [n] bodyimag [n]
  /TREE DISPLAY=TOPDOWN NODES=STATISTICS
   BRANCHSTATISTICS=YES NODEDEFS=YES SCALE=AUTO
  /DEPCATEGORIES USEVALUES=[VALID]
  /PRINT MODELSUMMARY CLASSIFICATION RISK
  /METHOD TYPE=CRT MAXSURROGATES=AUTO PRUNE=SE(1)
  /GROWTHLIMIT MAXDEPTH=8 MINPARENTSIZE=100 MINCHILDSIZE=50
  /VALIDATION TYPE=NONE OUTPUT=BOTHSAMPLES
  /CRT IMPURITY=GINI MINIMPROVEMENT=0.0001
  /COSTS EQUAL
  /PRIORS FROMDATA ADJUST=NO
  /MISSING NOMINALMISSING=MISSING.
```

Using Classification and Regression Trees, page 147

APPENDIX H

SPSS Decision Tree Syntax for Mental Health and Physical Health Data

```
TREE mentheal [s] BY physheal [s] sex [n] age [s] fses [s]
mses [s] numpar [s] FORCE=physheal [s]
  /TREE DISPLAY=TOPDOWN NODES=STATISTICS
   BRANCHSTATISTICS=YES NODEDEFS=YES SCALE=AUTO
  /PRINT MODELSUMMARY RISK
  /GAIN SUMMARYTABLE=YES TYPE=[NODE] SORT=DESCENDING
   CUMULATIVE=NO
  /METHOD TYPE=CHAID
  /GROWTHLIMIT MAXDEPTH=AUTO MINPARENTSIZE=100
   MINCHILDSIZE=50
  /VALIDATION TYPE=SPLITSAMPLE(50.00) OUTPUT=TESTSAMPLE
  /CHAID ALPHASPLIT=0.05 ALPHAMERGE=0.05 SPLITMERGED=NO
   ADJUST=BONFERRONI INTERVALS=10
  /MISSING NOMINALMISSING=MISSING.
```

APPENDIX I

SPSS Decision Tree Syntax for CART Produce for Meta-Analysis

```
TREE ES [s] BY D1 [n] D2 [n] D3 [n] D4 [n] RDesign [n]
StuMeas [n] Minority [n]
  /TREE DISPLAY=TOPDOWN NODES=STATISTICS
   BRANCHSTATISTICS=YES NODEDEFS=YES SCALE=AUTO
  /PRINT MODELSUMMARY RISK
  /GAIN SUMMARYTABLE=YES TYPE=[NODE] SORT=DESCENDING
   CUMULATIVE=NO
  /METHOD TYPE=CRT MAXSURROGATES=AUTO PRUNE=NONE
  /GROWTHLIMIT MAXDEPTH=4 MINPARENTSIZE=10 MINCHILDSIZE=5
  /VALIDATION TYPE=CROSSVALIDATION(10) OUTPUT=BOTHSAMPLES
  /CRT MINIMPROVEMENT=0.0001
  /INFLUENCE w
  /MISSING NOMINALMISSING=MISSING.
```

Using Classification and Regression Trees, page 151

About the Author

Dr. **Xin Ma** is a full professor of quantitative and psychometric methods and mathematics education at the University of Kentucky. He earned his doctoral degree from the University of British Columbia in Canada. He is a Spencer (postdoctoral) Fellow of the (U.S.) National Academy of Education. He was a former Canada research chair and the former founder and director of the Canadian Centre for Advanced Studies of National Databases at the University of Alberta.

Dr. Ma has been teaching various statistics courses at the graduate level for more than 20 years. His main research interests include advanced quantitative methods, large-scale data analysis, and mathematics education. He is the author of the book *A National Assessment of Mathematics Participation in the United States: A Survival Analysis Model for Describing Students' Academic Careers* (1997, Edwin Mellen). He has published numerous refereed articles in prestigious academic journals such as *American Educational Research Journal, American Journal of Education, Educational Evaluation and Policy Analysis, Journal for Research in Mathematics Education,* and *Teachers College Record.*

Using Classification and Regression Trees, page 153

Printed and bound by CPI Group (UK) Ltd, Croydon, CR0 4YY
15/09/2026
14962335-0008